The text is a self-contained, comprehensive introduction to the theory of hydrodynamic lattice gases.

Lattice-gas cellular automata are discrete models of fluids. Identical particles hop from site to site on a regular lattice, obeying simple conservative scattering rules when they collide. Remarkably, at a scale larger than the lattice spacing, these discrete models simulate the Navier-Stokes equations of fluid mechanics. This book addresses three important aspects of lattice gases. First, it shows how such simple idealized microscopic dynamics give rise to isotropic macroscopic hydrodynamics. Second, it details how the simplicity of the lattice gas provides for equally simple models of fluid phase separation, hydrodynamic interfaces, and multiphase flow. Last, it illustrates how lattice-gas models and related lattice-Boltzmann methods have been used to solve problems in applications as diverse as flow through porous media, phase separation, and interface dynamics. Many exercises and references are included.

This book should be of interest to physicists, mathematicians, engineers, earth scientists, and computer scientists interested in the simulation of fluid dynamics.

COLLECTION ALEA-SACLAY:
MONOGRAPHS AND TESTS IN STATISTICAL PHYSICS

General Editor: Claude Godrèche

LATTICE-GAS CELLULAR AUTOMATA

Lattice-Gas Cellular Automata

Simple models of complex hydrodynamics

DANIEL H. ROTHMAN

Massachusetts Institute of Technology

STEPHANE ZALESKI

Université Pierre et Marie Curie, Paris

PUBLISHED BY THE PRESS SYNDICATE OF THE UNIVERSITY OF CAMBRIDGE
The Pitt Building, Trumpington Street, Cambridge, United Kingdom

CAMBRIDGE UNIVERSITY PRESS
The Edinburgh Building, Cambridge CB2 2RU, UK
40 West 20th Street, New York NY 10011–4211, USA
477 Williamstown Road, Port Melbourne, VIC 3207, Australia
Ruiz de Alarcón 13, 28014 Madrid, Spain
Dock House, The Waterfront, Cape Town 8001, South Africa

http://www.cambridge.org

First published 1997
First paperback edition 2004

Typeset in 10/13 Computer Modern

A catalogue record for this book is available from the British Library

Library of Congress Cataloguing in Publication data

Rothman, Daniel H.
Lattice-gas cellular automata: simple models of complex hydrodynamics / Daniel H. Rothman,
Stéphane Zaleski.
p. cm. – (Collection Aléa-Saclay; 5)
includes bibliographical references and index.
ISBN 0 521 55201 X hardback
1. Hydrodynamics–Mathematical models. 2. Hydrodynamics–Computer simulation.
3. Lattice gas–Mathematical models. 4. Cellular automata–Mathematical models.
I. Zaleski, S. II. Title. III. Series.
QC157.R587 1997
532.05015113dc21   96-71777 CIP

ISBN 0 521 55201 X hardback
ISBN 0 521 60760 4 paperback

To our parents

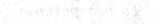

Contents

Preface

We believe that the computer simulation of physics should retain the elegance of physics itself. This book is about one approach towards this objective: *lattice-gas cellular automata* models of fluids.

Cellular automata are fully discrete models of physical and other systems. Lattice gases are just what the name says: a model of a gas on a grid. Thus lattice-gas cellular automata are a special kind of gas in which identical particles hop from site to site on a lattice at each tick of a clock. When particles meet they collide, but they always stay on the grid and appropriate physical quantities are always conserved.

Our subject is interesting because it provides a new way of thinking about the simulation of fluids. It also provides an instructive link between the microscopic world of molecular dynamics and the macroscopic world of fluid mechanics. Lastly, it allows us to create new computational tools that can be usefully applied to solve certain problems.

There are, broadly speaking, two ways to use computers to make progress in physics. The first approach is to use computers to compute a number, say the result of a certain integral or the hydrodynamic drag past a certain body. The second is to use computers as a kind of experimental laboratory, to explore the phenomena of interest much as an experimentalist would do him or herself. In the former case, realism is essential—if you do not solve the right equations, then you will not compute the right drag. In the latter case, however, simplicity is equally valued, since it greatly aids the identification and exploration of the system. Because cellular automata are highly idealized models of physics, they are usually identified with the second approach.

Lattice gases, however, bring something new to cellular automata: they are not only simple, but they are also a realistic model of the equations of hydrodynamics. In fact, they provide a simple framework for the development of a kinetic theory and the consequent derivation of hydrodynamic equations. Naively, one would think that it would be impossible to model isotropic fluid mechanics by constructing a fictitious molecular dynamics in which the molecules all move at the same speed and in only a limited number of directions. It turns out to be not only possible, however, but also the remarkably simple consequence of geometric symmetry and physical conservation laws. Why this is so is one of the three principal themes of this book.

The second theme is the discrete modeling of immiscible fluid mixtures. The simplicity of the lattice gas allows for equally simple models of fluid phase separation and hydrodynamic interfaces. The behavior of these models is striking, because they not only include within them the macroscopic hydrodynamics of the plain lattice gas, but also aspects of the statistical mechanics of phase transitions and interface fluctuations. This is somewhat surprising and mysterious because these models are microscopically irreversible, and thus may be more representative of dynamical systems than thermal systems. Though we have much insight to report, a complete and fundamental understanding of these models remains elusive.

The third theme of this book amounts to an attempt to answer the question, "What are lattice gases good for?" We adopt here a pragmatic view—rather than evaluating the efficiency of the method or detailing what problems may be simulated, we concentrate instead on some of the problems that lattice-gas simulations have *solved*. Thus we focus on a few of the areas of fluid physics—flows through porous media, phase separation, and interface dynamics—where lattice gases have made significant contributions.

We begin with a brief recounting of the intellectual history of the subject. Lattice-gas automata are in some sense a consequence of the early visions of von Neumann and Ulam of a world in which computers could act as a kind of mathematical laboratory. They are also, insofar as physics is concerned, the most impressive outcome of the considerable effort that physicists, computer scientists, and others have made towards the construction of simple physical models—i.e., cellular automata—that may be precisely stated in terms of Boolean logic.

The reader soon finds out that the lattice gas is also of interest not only to physics and computer science but also to applied mathematics, geophysics, and those branches of engineering that concern themselves with multiphase and complex fluids. Thus as authors we cannot expect that the two core subjects of our field—hydrodynamics and statistical mechanics—are already known to our readers. We therefore present in Chapter 2 a review of some of the fundamentals of fluid mechanics, and later, in Chapter 14, an overview of some relevant issues in statistical mechanics. The material in Chapter 2 acts as the point of departure for almost all that follows, in that it contrasts the classical continuum picture of fluids to the discrete particle view of the lattice gas. It also contains an elementary derivation of the hydrodynamic equations of the lattice gas, in addition to establishing many of the notational conventions.

Chapters 3 and 4 present detailed derivations of the Euler equation and viscosity coefficient of two-dimensional lattice gases. Our presentation is pedagogical, allowing the reader to learn how macrodynamical equations may be quantitatively derived from microscopic dynamics. We follow these developments with a discussion of three-dimensional models in Chapter 5. Here our discussion is brief: we emphasize geometry, symmetry, and the principles upon which collision rules are designed. We delay a derivation of the complete viscous (Navier-Stokes) equations of D-dimensional lattice gases until Chapter 15.

Whereas lattice gases are *models* of fluids, concepts from the world of lattice gases may be used to construct *lattice-Boltzmann methods*, which are tools for the numerical simulation of fluid dynamics. The lattice-Boltzmann method has useful stability properties in addition to allowing easy vectorization and parallelization, thus making it an attractive alternative to finite-difference and finite-element methods. Our discussion of Boltzmann methods in Chapters 6 and 7 serves as an introduction to this subject.

We embark upon the subject of fluid mixtures with a discussion of miscible fluids in Chapter 8, and then move immediately to models of immiscible fluids. As we have already mentioned, the models of immiscible fluids are central to this book. Not only are they of interest for statistical mechanics, but they also may be used as tools, particularly in the case of the associated lattice-Boltzmann algorithms, for the simulations of certain multiphase flows. Chapters 9 through 12 are an exposition of some of the main aspects of this subject.

One of the most successful applications of the lattice gas has been to problems of flow through porous media. Thus Chapter 13 is dedicated entirely to this subject. We emphasize the problems that have been solved, and hope to impart the intrinsic interest of the subject.

Our goals are similar in the discussions of phase separation and interfaces that are the subjects of Chapters 16 and 17. The latter case, in particular, has recently led to some remarkable results concerning the fluctuating hydrodynamics of lattice-gas interfaces.

In the final chapter we briefly survey what is probably properly termed one of the vanguards of the field: complex fluids and patterns. The models we discuss there all produce beautiful images. But we hope that the pretty pictures do not detract from their scientific interest! Indeed, particularly in the case of the models of three-phase flow and microemulsions, we expect substantial further activity in the near future.

Having detailed the organization of the book, it is also worthwhile to

point out a few of its stylistic features. Not long ago we wrote a long review paper. The form of our review fell well within the genre: summaries
of our colleagues' (and our own!) work, accompanied by the requisite
extensive citations. We liked the product, but we felt that the reliance
upon references to "fill in the gaps" was inappropriate for a book intended
to reach a wider audience, i.e., a readership with a background similar
to that of a beginning graduate student in physical science or engineering. Thus in this book we have attempted to avoid citations as much
as possible within the text, but we follow each chapter with appropriate
bibliographic, historical, and technical notes. We have found that this
style has forced upon us a kind of self-contained clarity. (It also created
a lot of work!) The result is that there is an overlap of probably only
10–15% between our review and this book, plus a number of new results.
We hope that our effort has made the subject far more accessible. We
also hope that our colleagues will view increased accessibility as a benefit
to be reaped, because it has come at the cost of giving only the mininum
number of bibliographic citations.

Space and time limitations have made it impossible for us to provide an
adequate introduction to the practical aspects of the computer codes for
lattice-gas models and lattice-Boltzmann methods. Nevertheless we have
assembled many of our codes at a site accessible from the World Wide
Web.* Many of the codes are written in a pedagogical style, and some
have already been used in a graduate course at MIT. They are available
principally to give readers an idea of how some of the algorithms may
be written, rather than as simulation tools ready to be applied to new
problems. We hope that they will find some use.

Cambridge and Paris   Daniel H. Rothman
March, 1997   Stéphane Zaleski

---

* At the time of writing, this site is located at `ftp://ftp.jussieu.fr/jussieu/labos/lmm`.
  Interested readers of the future should seek out appropriate pointers at the web site of
  Cambridge University Press.

Acknowledgements

We would like to first thank our doctoral students whose work has led to results described in this book. At MIT, Andrew Gunstensen's dissertation gave rise to much of Chapter 10, while John Olson's thesis was responsible in large part for the material of Chapters 11 and 13. In Paris, the work of Cécile Appert and Valérie Pot formed the basis of Chapter 12, and Cécile later performed much of the work described in Chapter 16. The work of Bruno Lafaurie inspired the stability analysis in Chapter 7.

Much of this book has served as material for graduate courses at both MIT and the University of Paris. We thank the students of these courses for their patience with non-existent or half-written notes. In particular we thank Olav van Genabeek at MIT for his help in preparing the computer codes used in this course, some of which may be found at the web site mentioned in the preface. Olav also kindly prepared Figures 6.1 and 8.2.

Collaborations with our colleagues have been indispensable. Dominique d'Humières contributed substantially to the theoretical understanding of interfaces described in Sections 10.2 and 17.1, in addition to providing his expertise and insight in innumerable other ways. The work on interface fluctuations discussed in Sections 17.4 and 17.5 resulted from a fruitful collaboration with Eirik Flekkøy. Eirik provided invaluable suggestions for these two sections and Appendix F. He also provided the figures from his own work discussed in Chapter 8.

Discussions with Bruce Boghosian concerning his work on microemulsions helped make Section 18.5 possible. Bruce also kindly provided Figure 18.6.

Mathieu Ernst provided an extensive critique of some earlier work of ours, and many of his suggestions have worked their way into this book.

Interactions with Leo Kadanoff and Yves Pomeau over a number of years have led us to a better appreciation and understanding of our subject, much of which has intellectual roots in their own work. Both of us have profited greatly from many visits to the University of Chicago to discuss our work with Leo and other colleagues (mentioned below). During one of these visits, Leo collaborated with DHR on the work illustrated by Figure 18.5. During another, SZ was able to make invaluable progress on the liquid-gas model of Chapter 12.

Numerous other colleagues and students have either collaborated with us or were in some other way helpful in our work. Among them, we mention Chris Adler, Bernie Alder, Jean Pierre Boon, Roland Borghi, Christopher Burges, Henri Cabannes, Shiyi Chen, Paul Clavin, Peter Constantin, Liliana di Pietro, Gary Doolen, Jens Feder, Bruno Ferréol, Uriel Frisch, Renée Gatignol, Laurent Giraud, Denis Gueyffier, Howard Gutowitz, Michel Hénon, Richard Holme, Claude Jaupart, Dominique Jeulin, Ray Kapral, Jeff Keller, Pierre Lallemand, Anna Lawniczak, Joel Lebowitz, Norm Margolus, Nick Martys, Terizhandur Ramakrishnan, Frank Richter, Alastair Rucklidge, Ralph Santos, Larry Schwartz, Geoff Searby, Wendy Soll, Jon Somers, Sauro Succi, Albert Tarantola, Tom Toffoli, Gerard Vichniac, Tom Witten, and Gianluigi Zanetti.

Many of our figures have been published previously, and we are grateful to each of the original copyright holders for their kind permission to reprint these figures. Whereas the original publication is always appropriately cited, a few of our figures require more formal acknowledgment. These are:

- Figure 13.3, reprinted from Figure 2 of Ref. [13.6], copyright 1995 Kluwer Academic Publishers.

- Figure 18.2, reprinted from Figures 3 and 4 of Ref. [18.2], copyright 1991 Elsevier Science - NL, Sara Burgerhartstraat 25, 1055 KV Amsterdam, The Netherlands.

- Figures 18.4 and 18.5, reprinted from Figures 1 and 2 of Ref. [18.5], copyright 1994 American Institute of Physics.

- Figure 18.6, reprinted from Figures 11 and 13 of Ref. [18.7], copyright 1996 The Royal Society.

Most of the writing of this book and much of the work that preceded it was done in Paris. DHR would like to thank the Centre National de Recherche Scientifique (CNRS) and the Laboratoire de Physique Statistique of the École Normale Supérieure for support in 1992–93. Additional funding, at that time and later, was provided by the École des Mines de Paris, École Normale Supérieure, Institut de Physique du Globe de Paris, Institut Universitaire de France (IUF), and NATO.

SZ would like to thank the CNCPST, Direction des Recherches, Études et Techniques, IDRIS, and IUF. Additional support was provided by the

CNRS, mainly through the Groupements de Recherche "Mécanique des Fluides Numérique" and "Moteurs Fusées". He would also like to thank Jeff Yepez for arranging U. S. Air Force funding for a crucial trip to MIT.

In this time of decreasing governmental support for basic research, it is a pleasure for DHR to express his gratitude to Art Thompson of Exxon and Alain Labastie of Elf Aquitaine for their continued industrial support. Elf also funded the work of Bruno Ferréol at MIT, which led to some of the results of Chapter 13. When the work that led to this book started in the late 1980's, several other oil companies provided generous support. DHR would like to thank Sven Treitel and Lee Baker of Amoco and Ernest Chung of Chevron for making this industrial funding happen. At various times, Agip, BP, Mobil, and Shell also provided some support. Additional funding for the work at MIT has also come from the National Science Foundation and the Petroleum Research Fund, administered by the American Chemical Society.

Finally, we would like to express our gratitude for the patience, understanding, and love of our wives, Claude and Isabelle. SZ also hopes his children will someday understand why he put so much time into this book.

1

A simple model of
fluid mechanics

Although the microscopic makeup of fluids ranges from the simplest mon-atomic gas to, say, a complex mixture such as milk, nearly all fluids flow in a way that obeys the same equations of fluid mechanics. How simple can a model of a fluid be and still satisfy these same equations?

In this chapter we introduce a microscopic model of a fluid that is far simpler than any natural fluid. Indeed, it has nearly nothing in common with real fluids except for one special property—at a macroscopic scale it flows just like them!

This simple model represents an attempt to digitize, or reduce to logic, the equations of motion of hydrodynamics. After a discussion of the model's historical relation to other such attempts to simplify physics to make it more adaptable to computation, we consider some of the specific ramifications of the discovery of this simple fluid. This chapter establishes the context for the more detailed analyses, extensions, and applications of this model that follow.

1.1   The lattice gas

In 1986, Uriel Frisch, Brosl Hasslacher, and Yves Pomeau announced a striking discovery. They showed that the molecular, or atomistic, motion within fluids—an extraordinarily complicated affair involving on the order of 10^{24} real-valued degrees of freedom—need not be nearly so detailed as real molecular dynamics in order to give rise to realistic fluid mechanics. Instead, a fluid may be constructed from fictitious particles, each with the same mass and moving with the same speed, and differing only in their velocities. Moreover, the velocities themselves are restricted to a finite set. Indeed, to construct a two-dimensional fluid, one needs only six!

The particles of this simple fluid live on a hexagonal lattice, as shown in Figure 1.1. The laws of motion that these particles follow are easy

1

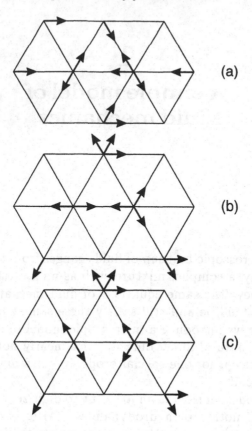

Fig. 1.1. An example of one time step in the evolution of a model of a simple fluid [1.19]. Each arrow represents a particle of unit mass moving in the direction given by the arrow. Figure 1.1a is the initial condition. Figure 1.1b represents the propagation, or free-streaming step: each particle has moved one lattice unit in the direction of its velocity. Figure 1.1c shows the result of collisions. The only collisions that have changed the configuration of particles are located in the middle row.

to describe. First, the lattice is initially prepared so that no more than one particle is moving with a particular velocity at a particular site on the lattice (Figure 1.1a). Then, at each tick of a clock, the particles *hop* and *scatter*. The hopping step is illustrated by Figure 1.1b: each particle moves one lattice unit in the direction of its velocity. Scattering is shown in Figure 1.1c: if two or more particles arrive at the same site, they can collide. Some collisions cause the particles to scatter—that is, their velocities change—while others do not. In all cases, however, collisions may change neither the total number of particles nor the vector sum of their velocities. In other words, mass and momentum are conserved.

This model is called a *lattice gas*. As we discuss further below, it differs from previous lattice gases in that momentum is explicitly conserved. Indeed, the conservation of momentum in conjunction with the conservation of mass allows one to write equations that describe the evolution of these quantities in the model. At the microscopic scale of the lattice, these equations are just a formalization of the dynamics pictured in Figure 1.1. Surprisingly, however, coarse-grained averages of these microdynamical equations can be shown to be asymptotically equivalent to

$$\nabla \cdot \mathbf{u} = 0 \tag{1.1}$$

and

$$\frac{\partial \mathbf{u}}{\partial t} + \mathbf{u} \cdot \nabla \mathbf{u} = -\frac{1}{\rho}\nabla p + \nu\nabla^2\mathbf{u}. \tag{1.2}$$

These are the *Navier-Stokes equations* of incompressible hydrodynamics. Here $\mathbf{u}$ and p are a velocity and pressure, respectively, that may be defined from coarse-grained averages of the particle motion, ρ represents a coarse-grained particle density, and time derivatives and gradients are taken at scales much slower than a time step and much greater than a lattice unit, respectively. The viscosity ν depends on the precise details of the collision rule.

Equation (1.1) is the statement of mass conservation in an incompressible fluid. Similarly, equation (1.2) is a statement of momentum conservation, or, more specifically, Newton's second law, which, in this case, equates the acceleration of a macroscopic piece of fluid to external and internal stresses felt by this piece of fluid. One could naively argue that neither equation is a remarkable consequence of the particle dynamics we have just described—after all, mass and momentum conservation at a microscopic scale must still hold at any macroscopic scale. But such an argument misses a truly amazing result: not only do we have hydrodynamic equations at a macroscopic scale, but these equations are *isotropic*. In other words, despite the fact that the microscopic particles can move in only six possible directions, the two-dimensional macroscopic fluid that they compose has no preferred directions of flow—just like a real fluid!

In Chapter 2 we derive equation (1.1) and the inviscid form of equation (1.2), first at the level of continuum mechanics and then from the discrete particle dynamics of the lattice gas. The latter derivation will be pedagogical but crude. A rigorous derivation of the hydrodynamic limit of the lattice gas requires considerable work, so we delay it until later chapters. To convince you now, however, that the lattice gas can simulate hydrodynamics, we appeal first to a picture, and then to what is probably the simplest possible verbal statement of the physics of the lattice gas.

The picture is Figure 1.2; it shows a flow simulated with the lattice gas. This result is from a series of the first simulations to be performed with

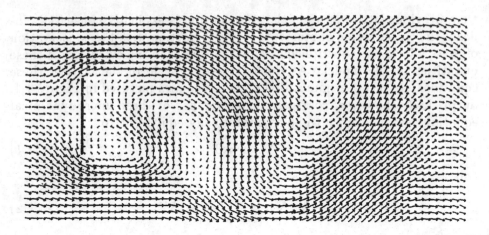

Fig. 1.2. Two-dimensional flow past a flat plate, simulated using the lattice gas [1.6]. The arrows represent the direction of flow, which is forced from left to right. Each arrow was computed from the average velocity in a cell containing 16 × 16 sites on a hexagonal lattice.

the lattice gas, and we show it more for its historical interest rather than to indicate the present "state of the art." This particular flow is forced from left to right. The Reynolds number, a dimensionless number that indicates the degree of turbulence, is about 70, high enough to result in the unsteady vortex shedding, known as *Bénard-von Kármán streets*, behind the flat plate. This simulation qualitatively matches the same results that one would expect from a nearly two-dimensional flow (as seen, for example, when a flag blows gently in the wind) or from an equivalent simulation performed with a different numerical method.

Such early simulations of the lattice gas captured the attention of many scientists, among them the physicist Richard Feynman. Noticing that his colleagues were having trouble explaining the lattice gas to others, he suggested that they explain it as follows:

> We have noticed in nature that the behavior of a fluid depends very little on the nature of the individual particles in that fluid. For example, the flow of sand is very similar to the flow of water or the flow of a pile of ball bearings. We have therefore taken advantage of this fact to invent a type of imaginary particle that is especially simple for us to simulate. This particle is a perfect ball bearing that can move at a single speed in one of six directions. The flow of these particles on a large enough scale is very similar to the flow of natural fluids.

Feynman's interest in the lattice gas derived in part from an interest in whether computers could be used to perform an *exact* simulation of physics, rather than the approximation one normally obtains from numerical solutions to differential equations. At the heart of such an idea is the subject of *cellular automata*, to which we now turn.

1.2 Cellular automata

In the late 1940's, at a time when the potential of digital computing was just beginning to be considered, the mathematicians Stanislaw Ulam and John von Neumann proposed that computers could be programmed to simulate life. The idea was that one could place simple computing machines, each capable of representing a finite number of states, at each node of a lattice or graph. Each of the machines would be connected to some of its neighbors, and the states of the neighbors at time t_n would determine the state of the machine connected to them at time t_{n+1}. Ulam and von Neumann proposed that under certain conditions, the collective behavior of the entire system could be sufficiently complex so that the system could imitate aspects of real biological systems, specifically the ability to reproduce itself. Systems of interconnected, "finite-state" machines became known as cellular automata.

Applications of these ideas to physics and mathematics were also considered at the time. Ulam was perhaps the first to propose that one could use computers to perform a kind of analog physical experiment. Rather than attempting a numerical solution of a particular differential equation, he argued, it could instead be possible (and perhaps advantageous) to use a computer to simulate aspects of the underlying physics from which that equation could be derived. His intention was not to, say, show that an ensemble of random walks obeyed a diffusion equation, but to invert such a procedure. He posed the general problem in a lecture in 1950:

> Given a partial differential equation, we construct models of suitable games, and obtain distributions or solutions of the corresponding equations by playing these games, i. e., by experiment.

At about the same time, Konrad Zuse, a German engineer, considered some similar ideas and conceived of networks called *computing spaces* that could be applied to such problems.

In the 1980's, due partly to the advent of inexpensive bitmap graphics, the increasing ability of individual computer scientists to construct small but powerful computers in their own labs, and the growing realization that parallel computation would soon exert a powerful influence on the numerical simulation of physics in general and nonlinear physics in particular, these ideas of Ulam, von Neumann, Zuse, and others gained renewed importance. New hardware designed by Norman Margolus and Tommaso

Toffoli was particularly influential. They constructed the first inexpensive and widely useable *cellular automata machines*. Finally, one could not only construct theories about automata, but one could *see* them evolve on a computer screen. By this time, cellular automata were usually defined to be any discretely valued dynamical system that evolved in discrete time steps on a regular lattice by nearest neighbor interactions according to simple rules. Though one could embed virtually any spatially extended dynamical system into such a framework, the emphasis was on nonlinear systems in which the dynamical variables—that is, whatever quantity evolved according to the specified dynamics—could take on only a small, finite number of states. This aspect of the subject became of particular interest to Stephen Wolfram, who conducted a number of highly influential theoretical and computational studies of the simplest possible cellular automata that one could construct. By the mid-1980's he was among the most persuasive of a growing group of scientists who argued that a revolutionary advance in computational physics was forthcoming. The evidence was indeed compelling—some simple cellular automata that were known at the time behaved so much like real physical systems that there were predictions that they would someday replace, or at least supplement, existing methods of computer simulation. In particular, interest focused on the simulation of nonlinear physics. However, no cellular-automaton model for the simulation of a nonlinear partial-differential equation was yet known.

Or so it was thought. In the 1970's, Jean Hardy, Olivier de Pazzis, and Yves Pomeau had created a discrete model of a fluid on a square lattice. Each particle of this fluid moved with the same speed with only four possible velocities; the particles collided when they met in a way that conserved particle number and momentum. The model was shown to simulate hydrodynamic equations, but, to no one's surprise, the equations were anisotropic. The model lay dormant for about a decade until its connections with the growing interest in cellular automata were discovered. Soon afterward, Frisch, Hasslacher, and Pomeau realized that six velocities, rather than four, were necessary and sufficient to obtain isotropic hydrodynamics, and *lattice-gas cellular automata* were introduced as a model of the Navier-Stokes equations.

1.3   Ramifications

The introduction of the Frisch-Hasslacher-Pomeau lattice gas was accompanied by a tremendous amount of excitement. Aside from the intrinsic interest in the discovery that much of the details of real molecular dynamics were not necessary to create a continuum fluid, there was also immediate interest in the possibility that the lattice gas could be useful as a computationally efficient method for the numerical simulation of turbulent flows. This is a problem that has never lacked practical or economic interest, and the discovery of the lattice gas was even reported in the international press. A front-page article in the *Washington Post* broke the story with a statement that the lattice gas had the potential to be "one thousand to one million times faster than previous methods." At the time the Cold War still raged, and the *Post* also reported that the United States Department of Defense had considered whether the method "should be classified to keep it out of Soviet hands."

Fortunately there was never any secret. A large group of scientists and engineers soon became interested in determining precisely how closely the method simulated the Navier-Stokes equations and to what problems the method could be fruitfully applied.

First, it was confirmed that the macroscopic dynamics of the lattice gas were indeed very close to real two-dimensional hydrodynamics. Moreover a similar model for three-dimensional flow was soon found. There are deviations from the Navier-Stokes equation, but in all cases these are either insignificant for applications or they may be corrected by appropriate adjustments of the basic model.

Second, the computational efficiency of the model was addressed. As we show in later chapters, the lattice gas is indeed useful for certain problems, but the initial press reports were seriously exaggerated. It does remain possible, however, that computers in the future, particularly special-purpose cellular automata machines, could offer important advantages for lattice-gas methods.

Most important for the present, however, are some of the unexpected ramifications of the lattice gas that have emerged since its introduction. Specifically, the lattice gas has proven to be an interesting point of departure for the construction of models of certain complex fluids. We refer in particular to fluid mixtures including interfaces, exhibiting phase transitions, and allowing for multiphase flows. These models have an unusually rich phenomenology that derives in part from their ability to simulate phase-separation dynamics and fluid dynamics simultaneously. Moreover they may be usefully applied to some especially difficult practical problems in fluid mechanics, such as the simulation of multiphase flow through disordered media such as porous rocks. An explication of these models is indeed a principal objective of this book.

Now, one decade after the introduction of the lattice gas, one can point to an impressive list of achievements that have arisen from studies of the model itself and applications of it to problems in hydrodynamics and statistical mechanics. The most exciting applications of the lattice gas have usually exploited either the ease with which complex microphysics may be incorporated into the model (as in the case of interfaces) or the intrinsic fluctuations of the gas itself. These new results have usually been obtained from a combination of computational experiment and theoretical analysis. Before we describe them in the following chapters, however, it is useful to first remark on the role of the computer, and the lattice gas in particular, in physics.

1.4   A Boolean analog computer

Ulam's and von Neumann's interest in cellular automata arose in part from the conviction that computers should be used as heuristic aids to the theoretical understanding of nonlinear systems. In a 1946 lecture, von Neumann emphasized the need for progress in fluid dynamics in particular and mathematics in general:

> ...really efficient computing devices may, in the field of non-linear partial differential equations...provide us with those heuristic hints which are needed in all parts of mathematics for genuine progress. In the specific case of fluid dynamics these hints have not been forthcoming for the last two generations from the pure intuition of mathematicians. ... To the extent that such hints arose at all...they originated in a type of physical experimentation which is really computing. We can now make computing so much more efficient, fast and flexible that it should be possible to use the new computers to supply the needed heuristic hints. This should ultimately lead to important analytical advances.

To underline his point, von Neumann specifically stated that wind tunnels were analog computers ("computing devices of the so-called analogy type") for the integration of the Navier-Stokes equations. After all, he pointed out, the underlying physics of the system was not in doubt yet its behavior was not understood, making wind-tunnel experiments necessary for gaining fundamental insight.

By the same sort of reasoning, we see that the lattice gas is a kind of *Boolean* analog computer: rather than performing a real flow experiment, a simulation of it can be reduced to logic. To the extent that these logical expressions can be efficiently evaluated on a computer, a lattice-gas simulation may have a value commensurate with or exceeding that of a particular laboratory experiment. But this is not the only point. As with any analog computer, the behavior of the device itself can become of interest due to its simplicity and closeness to a particular physical problem. Nearly fifty years ago Ulam already recognized such a potential for an analog fluid, based on a "probability game," that he described conceptually:

...if one assumes as fundamental a model of [a] fluid as does the kinetic theory, the equations of hydrodynamics will describe the behavior of average quantities; velocities, pressures, etc., are defined by averaging these over very large numbers of atoms near a given position. The results of a probability game will reflect, to some extent, the deviation of such quantities from their average values. That is to say, the fluctuations unavoidably present as a result of the random processes performed may not be purely mathematical but may reflect, to some extent, the physical reality.

Thus we arrive at a fundamental point. As in Ulam's conceptual fluid, intrinsic fluctuations are an important and interesting property of the lattice gas. Although they make the lattice gas somewhat different from the Navier-Stokes equations, in many studies they add to, rather than detract from, the physical interest. Moreover, in some instances new aspects of hydrodynamics have been revealed as a consequence of these fluctuations. We shall see in the following chapters that fluctuations are but one of the many fascinating properties of this Boolean analog fluid.

1.5   Notes

The Frisch-Hasslacher-Pomeau lattice gas was introduced in:

1.1 Frisch, U., Hasslacher, B., and Pomeau, Y. (1986). Lattice-gas automata for the Navier-Stokes equations. *Phys. Rev. Lett.* **56**, 1505–1508.

1.2 Frisch, U., d'Humières, D., Hasslacher, B., Lallemand, P., Pomeau, Y., and Rivet, J.-P. (1987). Lattice gas hydrodynamics in two and three dimensions. *Complex Systems* **1**, 649–707.

Earlier discrete fluid models on a square lattice are in:

1.3 Kadanoff, L. P. and Swift, J. (1968). Transport coefficients near the critical point: a master equation approach. *Phys. Rev.* **165**, 310–322.

1.4 Hardy, J., Pomeau, Y., and de Pazzis, O. (1973). Time evolution of a two-dimensional model system. I. Invariant states and time correlation functions. *J. Math. Phys.* **14**, 1746–1759.

1.5 Hardy, J., de Pazzis, O., and Pomeau, Y. (1976). Molecular dynamics of a classical lattice gas: Transport properties and time correlation functions. *Phys. Rev. A.* **13**, 1949–1961.

The paper by Kadanoff and Swift is notable in part for its explicit study of a momentum-conserving gas in a system with a phase transition. It makes direct contact with the traditional lattice gas based on an Ising model, which has been used extensively to study the liquid-gas transition in the absence of hydrodynamics.

Figure 1.2 is from a series of simulations described in:

1.6 d'Humieres, D., Pomeau, Y., and Lallemand, P. (1985). Simulation d'allées de von Kármán bidimensionelles à l'aide d'un gaz sur réseau. *C. R. Acad. Sc. Paris* **301**, 1391–1394.

Feynman's interest in cellular automata is evident in:

1.7 Feynman, R. (1982). Simulating physics with computers. *Int. J. Theor. Phys.* **21**, 467–488.

1.8 Hillis, W. D. (1989). Richard Feynman and the Connection Machine. *Physics Today* **42**, February, 78–83.

The quotation from Feynman is on page 82 of the latter reference.

A lively historical account of Ulam's and von Neumann's self-reproducing automata is given in Chapter 18 of:

1.9 Dyson, F. (1979). *Disturbing the Universe* (Harper and Row, New York).

Zuse's work has been translated in:

1.10 Zuse, K. (1970). Calculating Space, MIT Project MAC, Tech. Transl. AZT-70-164-GEMIT.

Further historical comments on the history of cellular automata are given in:

1.11 Toffoli, T. and Margolus, N. (1987). *Cellular automata machines* (MIT Press, Cambridge).

Toffoli and Margolus also describe special-purpose hardware for cellular automata computations, in addition to presenting a wide phenomenology of the behavior of various models. Other aspects of cellular automata research in the 1980's can be found in:

1.12 Farmer, D., Toffoli, T., and Wolfram, S., editors (1984). *Cellular Automata* (North Holland Physics Publishing, New York).

1.13 Wolfram, S., editor (1986). *Theory and Applications of Cellular Automata* (World Scientific, Singapore).

1.14 Gutowitz, H., editor (1991). *Cellular Automata: Theory and Experiment* (MIT Press, Cambridge).

The quotations from Ulam are from pages 267–268 and page 270 of:

1.15 Ulam, S. (1952). Random Processes and Transformations. *Proceedings of the International Congress of Mathematicians* (held in 1950) **2**, 264–275.

Section VIII of Ulam's paper succinctly describes his and von Neumann's model of self-reproducing automata. Von Neumann's considerable efforts in this area are summarized and partly collected in:

1.16 von Neumann, J. (1966). *Theory of Self-reproducing Automata (edited and completed by A. Burks)* (University of Illinois Press, Urbana).

The quotation from von Neumann is on page 4 of:

1.17 Goldstine, H. H. and von Neumann, J. (1946). On the principles of large scale computing machines. In Taub, A. H., editor, *John von Neumann: Collected Works 5: Design of Computers, Theory of Automata, and Numerical Analysis* (Macmillan, New York, 1963).

The first news report of the lattice gas was in:

1.18 Hilts, P. J. (1985). Discovery in flow dynamics may aid car, plane design. *Washington Post*, November 19, 1.

This book represents an expansion of the point of view in:

1.19 Rothman, D. H. and Zaleski, S. (1994). Lattice-gas models of phase separation: interfaces, phase transitions, and multiphase flow. *Rev. Mod. Phys.* **66**, 1417–1479.

2
Two routes to hydrodynamics

Our objectives for this chapter are twofold. First, we review some elementary aspects of fluid mechanics. We include in that discussion a classical derivation of the Navier-Stokes equations from the conservation of mass and momentum in a continuum fluid. We then discuss the analogous conservation relations in a lattice gas. Finally, we briefly describe the derivation of hydrodynamic equations for the lattice gas, but defer our first detailed discussion of this subject to the following chapter.

2.1 Molecular dynamics versus continuum mechanics

The study of fluids typically proceeds in either of two ways. Either one begins at the microscopic scale of molecular interactions, or one assumes that at a particular macroscopic scale a fluid may be described as a smoothly varying continuum. The latter approach allows us to write conservation equations in the form of partial-differential equations. Before we do so, however, it is worthwhile to recall the basis of such a point of view.

The macroscopic description of fluids corresponds to our everyday experience of flows. Figure 2.1 shows that a flow may have several characteristic length scales l_i. These lengths scales may be related either to geometric properties of the flow such as channel width or the diameter of obstacles or to intrinsic properties such as the size of vortical structures. The smallest of these length scales will be called L_{hydro}. In some cases, such as low-velocity flow in a channel, L_{hydro} is simple to determine, while in other cases, such as turbulent flow, eddies may exist on a whole range of scales. In any case a smallest length scale exists for the macroscopic description.

The microscopic description involves atoms or molecules of fluid, and the characteristic length scale is the average distance ℓ_{mfp} a molecule travels between collisions, called the *mean free path*. In some cases, such

12

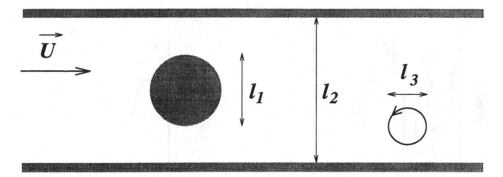

Fig. 2.1. Schematic diagram of a flow in a channel showing the typical characteristic length scales l_1, l_2, and l_3. The smallest hydrodynamic length scale L_{hydro} in such a flow must be much larger than the molecular length scale to allow a continuum description of the flow.

as liquids, ℓ_{mfp} may be very short, of the order of the distance L_1 between molecules, which is usually just a few angstroms. In other cases, such as dilute gases, ℓ_{mfp} may be much larger than L_1. In the discrete world of lattice gases, the lattice spacing introduces yet another microscopic length scale. In any case ℓ_{mfp} is an upper bound on microscopic length scales.

A basic hypothesis underlying continuum mechanics is that the continuum description holds whenever the characteristic length scale of macroscopic motions is much larger than the characteristic length scale of microscopic phenomena. In symbols,

$$L_{\text{hydro}} \gg \ell_{\text{mfp}}. \tag{2.1}$$

Whenever this condition holds, the evolution of the macroscopic field may be described by continuum mechanics, i.e., partial-differential equations.

Because these concepts are so important for understanding the transitions from the microscopic to the macroscopic world, it is worth looking at them in some more detail. A thought experiment may be used to get a feeling for microscopic and macroscopic length scales. Consider a device that measures density in a box of size ℓ while moving through a fluid. At each instant the experimenter counts the number of molecules in the box. To fix ideas, assume that the device moves a distance that is about ten times its size ℓ. At small length scales there are large fluctuations in the measured density as shown in Figure 2.2a. As the size ℓ increases the measured density fluctuates less. The relative amplitude of fluctuations is proportional to $N^{-1/2}$, where $N \sim \rho_n \ell^3$ is the number of molecules in the box and ρ_n is the average number of molecules per unit volume. The fluctuations become negligible when $\ell \gg L_1$. On the other hand, at scales of the order of L_{hydro}, we expect smooth variations

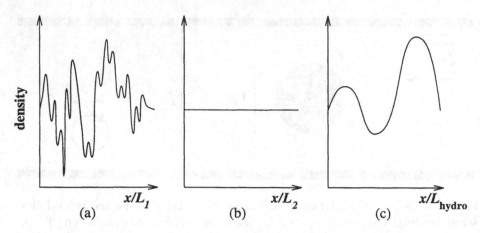

Fig. 2.2. Results of the thought experiment described in the text. In (a) the measuring device is of size $\ell \sim L_1$, the characteristic distance between molecules. Thus large noisy microscopic fluctuations are seen. In (c) the measuring device is of size $\ell \sim L_{\text{hydro}}$, and smooth macroscopic fluctuations are seen. In (b) the scale $\ell \sim L_2$, intermediate between L_1 and L_{hydro}. No fluctuations are seen in this case.

of the density due to macroscopic motions (that might be connected, for instance, to a descending dense plume) or external forces, as shown in Figure 2.2c. Equation (2.1) implies that there is an intermediate length scale L_2 between L_1 and L_{hydro} such that fluctuations are small, as shown in Figure 2.2b. The existence of the scale L_2 illustrates the notion of a "scale gap"—specifically, the large distance between L_1 and L_{hydro}.

These concepts do not by any means exhaust the issue of the foundations of continuum mechanics. However, one of the major motivations for the study of lattice gases is that they allow us to address hypotheses underlying continuum mechanics in a simple framework. As we shall soon see, quantities such as ℓ_{mfp} may be explicitly calculated for the lattice gas, allowing us to verify whether the scale separation (2.1) holds. Partial-differential equations may then be written for the macroscopic evolution of mass and momentum on the lattice. Before we do that, however, we first review the classical derivation of these continuum equations for the case of real fluids. We begin with the conservation of mass.

2.2   Mass conservation and the equation of continuity

Consider an arbitrary volume V that is fixed in space, as shown in Figure 2.3. The mass in the volume is

$$\int \rho \, dV, \tag{2.2}$$

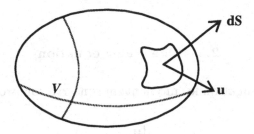

Fig. 2.3. A volume V of fluid. The surface element $\mathbf{dS}$ points outward and normal to the surface; $\mathbf{u}$ is the fluid velocity.

where ρ is the density and the integration is taken over the entire volume V. The fluid can move either into or out of the volume. Consider the motion through a surface element $\mathbf{dS}$, where $\mathbf{dS}$ is a vector normal to the surface, pointing outward, and of magnitude equal to the area of the surface element. The flow of fluid outward through $\mathbf{dS}$ is given by $\rho\mathbf{u}\cdot\mathbf{dS}$, where $\mathbf{u}$ is the fluid velocity. Thus the rate of mass loss from the volume V is then

$$\oint \rho\mathbf{u}\cdot\mathbf{dS},\tag{2.3}$$

where the integration is over the entire surface of the volume V. By taking the time derivative of expression (2.2), we find

$$\int \frac{\partial\rho}{\partial t}\,\mathrm{d}V = -\oint \rho\mathbf{u}\cdot\mathbf{dS},\tag{2.4}$$

where the negative sign on the right-hand side results from the outward sense of $\mathbf{dS}$. The surface integral may be converted to a volume integral via the divergence theorem:

$$\oint \rho\mathbf{u}\cdot\mathbf{dS} = \int \nabla\cdot\rho\mathbf{u}\,\mathrm{d}V.\tag{2.5}$$

Substitution of equation (2.5) into equation (2.4) then gives

$$\int \left(\frac{\partial\rho}{\partial t} + \nabla\cdot\rho\mathbf{u}\right)\,\mathrm{d}V = 0.\tag{2.6}$$

For this equation to be valid over any volume V, we must have

$$\frac{\partial\rho}{\partial t} + \nabla\cdot\rho\mathbf{u} = 0.\tag{2.7}$$

This is the *equation of continuity* that expresses conservation of mass in a fluid. In the case of *incompressible* fluids, the density may be taken to be approximately constant. In this case the continuity equation reduces to

$$\nabla\cdot\mathbf{u} = 0.\tag{2.8}$$

2.3   The Euler equation

Derivation of an equation for momentum conservation proceeds from Newton's second law,

$$\rho\frac{d\mathbf{u}}{dt} = \mathbf{F}, \tag{2.9}$$

where $\mathbf{F}$ is the force per unit volume acting on an element of fluid. We first consider the left-hand side. The total derivative $d\mathbf{u}/dt$ represents the acceleration of a "parcel" of fluid, such as the volume V in Figure 2.3, as it moves in space. The mass of this parcel is ρV, so the product $\rho d\mathbf{u}/dt$ is the rate of change of momentum for a parcel having unit volume. The total derivative, however, must be represented at a fixed point in space. There are two contributions to the derivative. First, the velocity can change at a fixed point simply because the velocity field itself may vary with time. This contribution is represented by the partial derivative $\partial\mathbf{u}/\partial t$. Second, velocity can be *advected* from one point in space to another by fluid motion. The two contributions sum such that a small change $d\mathbf{u}$ in the velocity results from a small change dt in time and small changes dx, dy, and dz in space according to

$$d\mathbf{u} = \frac{\partial\mathbf{u}}{\partial t}dt + \frac{\partial\mathbf{u}}{\partial x}dx + \frac{\partial\mathbf{u}}{\partial y}dy + \frac{\partial\mathbf{u}}{\partial z}dz. \tag{2.10}$$

By dividing both sides by dt and including once again the density we obtain

$$\frac{d\mathbf{u}}{dt} = \frac{\partial\mathbf{u}}{\partial t} + (\mathbf{u}\cdot\nabla)\mathbf{u}. \tag{2.11}$$

We must still account for the force density $\mathbf{F}$. There are two kinds of forces, those which are externally imposed and those which result from internal stresses. The former include body forces such as gravity and we do not bother to write them explicit. The latter include both pressure forces and forces that result from internal friction, or shearing stresses, in a flowing fluid. For the moment we assume that these shear stresses do not exist, or, in other words, that the fluid is inviscid. Thus we are left with only the pressure force acting on a fluid element, $-\oint p\,d\mathbf{S}$. Converting it to volume integral, we have

$$-\oint p\,d\mathbf{S} = -\int_{-\infty}^{\infty}\nabla p\,dV. \tag{2.12}$$

Thus the pressure force acting on a volume element of fluid is $-\nabla p\,dV$. After inserting it into the right-hand side of equation (2.9) and substituting (2.11), we arrive at the *Euler equation*

$$\frac{\partial \mathbf{u}}{\partial t} + (\mathbf{u} \cdot \nabla)\mathbf{u} = -\frac{1}{\rho}\nabla p. \qquad (2.13)$$

Equations (2.13) and (2.7) are a set of inviscid equations for compressible flow. They are not *closed*, however, since the pressure p is not specified. Several ways to close the equations through thermodynamical *equations of state* are indicated in Appendix C. The case of incompressible flow, on the other hand, is fully described by equations (2.13) and (2.8) and no further equations* are needed. Instead, in a solution of the equations, the pressure adjusts "automatically" to force the flow to follow the incompressibility condition (2.8). There is a trade-off involved when one describes a *weakly* compressible fluid, for instance water, as incompressible. While the need for equations of state is eliminated, the way in which pressure is related to the flow is somewhat more mysterious. This issue is one of the interesting aspects of classical continuum hydrodynamics, and is discussed at greater length in standard hydrodynamics textbooks such as those listed in the notes. Aspects of this problem will be made clearer in the lattice gas context, as we shall discover that the lattice gas is itself a weakly compressible fluid, for which we sometimes use the incompressible flow approximation.

2.4 Momentum flux density tensor

The Euler equation represents momentum conservation in an ideal inviscid fluid. Before moving on to consider the effects of viscosity, it is worthwhile to look at this momentum balance a little further. In this way we also have an opportunity to introduce some notation that will occur throughout this book.

We now consider the rate of change of momentum in a *fixed* volume, and therefore write

$$\partial_t(\rho u_\alpha) = \rho \partial_t u_\alpha + u_\alpha \partial_t \rho. \qquad (2.14)$$

Here u_α is the α-component of the velocity vector $\mathbf{u}$, and we have used the shorthand ∂_t for $\frac{\partial}{\partial t}$. By substitution of the Euler equation (2.13) into the first term on the right-hand side and the continuity equation (2.7) into the second term, we obtain

$$\partial_t(\rho u_\alpha) = -\rho u_\beta \partial_\beta u_\alpha - \partial_\alpha p - u_\alpha \partial_\beta(\rho u_\beta). \qquad (2.15)$$

---

* As usual for differential equations, we also need boundary and initial conditions for all the flow equations in this chapter.

Here we have used the Einstein summation convention in which repeated subscripts in a product imply a summation over those subscripts; i.e.,

$$X_\alpha Y_\alpha = \sum_{\alpha=1}^{D} X_\alpha Y_\alpha, \qquad (2.16)$$

where D is the number of space dimensions. Retaining this notation and combining derivatives, we can rewrite equation (2.15) as

$$\partial_t(\rho u_\alpha) = -\partial_\alpha p - \partial_\beta(\rho u_\alpha u_\beta). \qquad (2.17)$$

Finally, we obtain a simple expression for the momentum balance by writing

$$\partial_t(\rho u_\alpha) = -\partial_\beta \Pi_{\alpha\beta}^{(0)}, \qquad (2.18)$$

where we have introduced the inviscid *momentum flux density tensor*

$$\Pi_{\alpha\beta}^{(0)} = p\delta_{\alpha\beta} + \rho u_\alpha u_\beta \qquad (2.19)$$

via use of the Kronecker delta

$$\delta_{\alpha\beta} = \begin{cases} 1 & \alpha = \beta \\ 0 & \alpha \neq \beta. \end{cases} \qquad (2.20)$$

As shown in the exercises, $\Pi_{\alpha\beta}^{(0)}$ gives the flux of the α-component of momentum through a plane surface of unit area perpendicular to the β-direction. It is a tensor because it describes how a vectorial quantity, momentum, flows in a certain direction. Its inviscid form given by equation (2.19) is however inadequate for the description of momentum flux in real, and therefore viscous, fluids, to which we now turn.

2.5   Viscous flow and the Navier-Stokes equation

To describe internal stresses in viscous fluids, $\Pi_{\alpha\beta}^{(0)}$ must be modified to account for stresses due to shear and compression. The meaning of shear viscosity is illustrated by Figure 2.4, where a flow is divided by an imaginary plane positioned parallel to the fluid motion such that the velocity of the flow is faster above the plane than below it. Due to random motion of the molecules in a fluid, some of the molecules on the fast side of the plane cross to the slower side below, and, likewise, some of the "slow" molecules cross to the fast side above. As a result of this transfer of momentum, the fast side becomes slower and the slow side becomes faster. In a *Newtonian* fluid the momentum flux across the plane is proportional to the velocity gradient. The momentum flux due to shear for the specific case shown in Figure 2.4 is then

$$\Pi_{xy}^{\text{visc}} = -\mu \partial_y u_x, \qquad (2.21)$$

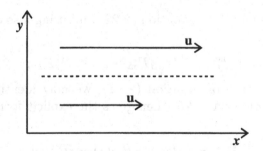

Fig. 2.4.   A simple shear flow $u_x \propto y$. Momentum transfer across the dotted line acts to damp out the velocity gradient.

where μ, the *dynamic shear viscosity*, is a positive material constant.

To describe shearing stresses for arbitrary velocity gradients in three dimensions, we must relate $\Pi_{\alpha\beta}^{\mathrm{visc}}$, the *viscous stress tensor*, to the rate of strain $E_{\alpha\beta}$ (defined below). For a Newtonian fluid, the two dimensional description of equation (2.21) is generalized by writing

$$\Pi_{\alpha\beta}^{\mathrm{visc}} = -\tilde{\mu}_{\alpha\beta\gamma\delta} E_{\gamma\delta}, \tag{2.22}$$

where $\tilde{\mu}_{\alpha\beta\gamma\delta}$ is a fourth-rank tensor that relates viscous stress to rate of strain, and we have implicitly included the possibility of compressional stress in addition to shear. Because equation (2.22) must be invariant under rotation or translation of the frame of reference, $\tilde{\mu}_{\alpha\beta\gamma\delta}$ must be an *isotropic* tensor. The most general form it may have is then

$$\tilde{\mu}_{\alpha\beta\gamma\delta} = A\delta_{\alpha\beta}\delta_{\gamma\delta} + B\delta_{\alpha\gamma}\delta_{\beta\delta} + C\delta_{\alpha\delta}\delta_{\beta\gamma}, \tag{2.23}$$

where A, B, and C are arbitrary constants.[†]

To determine $\Pi_{\alpha\beta}^{\mathrm{visc}}$ explicitly, it remains to define the strain rate $E_{\alpha\beta}$. Since $E_{\alpha\beta}$ is a measure of the deformation of a fluid element, it is clearly related to the velocity gradient $g_{\alpha\beta} = \partial_\alpha u_\beta$, and, by the assumption of Newtonian fluids, it must be a linear function of $g_{\alpha\beta}$. Note that we may write $g_{\alpha\beta}$ as a sum of symmetrical and asymmetrical terms:

$$g_{\alpha\beta} = \frac{1}{2}(g_{\alpha\beta} + g_{\beta\alpha}) + \frac{1}{2}(g_{\alpha\beta} - g_{\beta\alpha}). \tag{2.24}$$

The second term on the right-hand side corresponds to rigid-body rotation (or vorticity) and is not related to deformation. Thus we are left with only the symmetrical combinations of $g_{\alpha\beta}$, and we write

$$E_{\alpha\beta} = \partial_\alpha u_\beta + \partial_\beta u_\alpha. \tag{2.25}$$

[†] An isotropic tensor is a tensor whose components are unchanged by an orthogonal transformation (i.e., rotations and reflections) of coordinates. The derivation of equation (2.23) and related aspects of tensor symmetries are discussed in detail in Appendix A.

Substituting this form into equation (2.22) and using also equation (2.23), we obtain

$$\Pi_{\alpha\beta}^{\text{visc}} = -A\delta_{\alpha\beta}E_{\delta\delta} - (B + C)E_{\alpha\beta}. \tag{2.26}$$

From the special case of equation (2.21), we may identify $B + C = \mu$. Substitution of equation (2.25) then gives the explicit form of the viscous stress tensor,

$$\Pi_{\alpha\beta}^{\text{visc}} = -\mu(\partial_\beta u_\alpha + \partial_\alpha u_\beta) - \xi\partial_\gamma u_\gamma \delta_{\alpha\beta}, \tag{2.27}$$

where the last term on the right-hand side is related to compression, and $\xi = 2A$ is a positive coefficient related to the compression, or bulk, viscosity.

The momentum flux density tensor for viscous fluids is the sum of equations (2.19) and (2.27),

$$\Pi_{\alpha\beta} = \Pi_{\alpha\beta}^{(0)} + \Pi_{\alpha\beta}^{\text{visc}}, \tag{2.28}$$

and the resulting momentum-balance equation is

$$\partial_t(\rho u_\alpha) = -\partial_\beta \Pi_{\alpha\beta}. \tag{2.29}$$

Substitution of equations (2.19), (2.27), and (2.28) then gives

$$\partial_t(\rho u_\alpha) + \partial_\beta \rho u_\alpha u_\beta = -\partial_\alpha p + \partial_\beta[\mu(\partial_\beta u_\alpha + \partial_\alpha u_\beta)] + \partial_\alpha(\xi\partial_\gamma u_\gamma). \tag{2.30}$$

Equation (2.30) describes momentum conservation in a compressible viscous fluid. Though rather complicated, it may be considerably simplified for incompressible fluids. In this case we may take ρ to be constant and by equation (2.8), we can eliminate all terms involving the divergence $\partial_\alpha u_\alpha$. We also take the viscosity μ to be constant and use instead the *kinematic shear viscosity*

$$\nu = \mu/\rho. \tag{2.31}$$

We then obtain, in vector form,

$$\frac{\partial \mathbf{u}}{\partial t} + \mathbf{u} \cdot \nabla\mathbf{u} = -\frac{1}{\rho}\nabla p + \nu\nabla^2\mathbf{u}. \tag{2.32}$$

Equation (2.32) is the *Navier-Stokes equation* of incompressible viscous hydrodynamics.

2.6   Microdynamical equations of the lattice gas

We now turn to a heuristic derivation of the hydrodynamic equations for the lattice gas. We follow a similar path by writing the evolution equations for conserved quantities. However, instead of starting at the

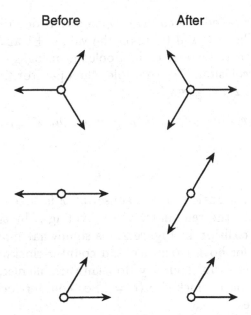

Fig. 2.5. Explicit examples of some collisions that may occur in the lattice gas on a hexagonal lattice [1.19]. The two-body head-on collision may result in either a clockwise or counter-clockwise rotation; here we show just one example. The two-body collision shown with non-zero net momentum results in no change, since no other configuration exists that conserves both the number of particles and the net momentum.

level of smoothly varying fields, we begin by writing equations for the particle dynamics.

Consider once again the lattice gas of Figure 1.1. The essential features of the dynamics are free propagation of the particles followed by collisions. Some explicit examples of collisions are shown in Figure 2.5. As stated earlier, some collisions can change the particle configurations, while others cannot. However, they must all conserve mass and momentum.

The particle dynamics depicted in Figure 1.1 may be expressed by the following equation:

$$n_i(\mathbf{x} + \mathbf{c}_i, t + 1) = n_i(\mathbf{x}, t) + \Delta_i[\mathbf{n}(\mathbf{x}, t)]. \qquad (2.33)$$

Here the time t is integer-valued and the duration of a time step is taken to be unity. The quantities $\mathbf{n} = (n_1, n_2, \ldots, n_6)$ are *Boolean* variables that indicate the presence ($n_i = 1$) or absence ($n_i = 0$) of particles moving from a lattice site situated at position $\mathbf{x}$ to the neighboring site situated at position $\mathbf{x} + \mathbf{c}_i$, where the particles move with unit speed in the directions given by

$$\mathbf{c}_i = (\cos \pi i/3, \sin \pi i/3), \qquad i = 1, 2, \ldots, 6. \qquad (2.34)$$

The function Δ_i is called the *collision operator*. It describes the change in $n_i(\mathbf{x}, t)$ due to collisions, and takes on the values ± 1 and 0. It is the sum of quantities that may be written as Boolean, or logical, expressions, one for each possible collision. For example, the operator for the three-body collision in Figure 2.5 is given by

$$\Delta_i^{(3)} = n_{i+1}n_{i+3}n_{i+5}\overline{n}_i\overline{n}_{i+2}\overline{n}_{i+4} - n_i n_{i+2} n_{i+4}\overline{n}_{i+1}\overline{n}_{i+3}\overline{n}_{i+5}, \quad (2.35)$$

where we have used

$$\overline{n}_i = 1 - n_i \quad (2.36)$$

as a shorthand for negation, and a subscript x is taken to imply "x mod 6"—in other words, the remainder after dividing x by six. The operator for the two-body collision in Figure 2.5 is somewhat more complicated if we wish to allow for both clockwise and counter-clockwise rotations. In this case we define a supplementary Boolean variable $a(\mathbf{x}, t)$ that indicates whether the collision is clockwise ($a = 1$) or counterclockwise ($a = 0$) so that we may write

$$\Delta_i^{(2)} = a n_{i+1} n_{i+4}\overline{n}_i\overline{n}_{i+2}\overline{n}_{i+3}\overline{n}_{i+5} + \overline{a} n_{i+2} n_{i+5}\overline{n}_i\overline{n}_{i+1}\overline{n}_{i+3}\overline{n}_{i+4} - n_i n_{i+3}\overline{n}_{i+1}\overline{n}_{i+2}\overline{n}_{i+4}\overline{n}_{i+5}. \quad (2.37)$$

For the simplest hexagonal lattice gas, the full collision operator Δ_i is

$$\Delta_i = \Delta_i^{(2)} + \Delta_i^{(3)}. \quad (2.38)$$

We refer to this combination of two-body and three-body collisions as the *FHP-I* model in deference to Frisch, Hasslacher, and Pomeau. More elaborate collision operators may be formed by including, for example, four-body collisions, or by allowing for collisions with particles at rest. Usually, the only restrictions on Δ_i are that it conserve mass,

$$\sum_i \Delta_i(\mathbf{n}) = 0, \quad (2.39)$$

and that it conserve momentum,

$$\sum_i \mathbf{c}_i \Delta_i(\mathbf{n}) = 0. \quad (2.40)$$

Using the first of these relations, one may sum the microdynamical equation (2.33) over each direction i to obtain an expression for the conservation of mass,

$$\sum_i n_i(\mathbf{x} + \mathbf{c}_i, t + 1) = \sum_i n_i(\mathbf{x}, t), \quad (2.41)$$

and, after multiplying the same equation by $\mathbf{c}_i$, summing again over i, and using the second relation, one obtains an expression for the conservation of momentum,

$$\sum_i \mathbf{c}_i n_i(\mathbf{x} + \mathbf{c}_i, t + 1) = \sum_i \mathbf{c}_i n_i(\mathbf{x}, t). \tag{2.42}$$

Equations (2.41) and (2.42) describe the evolution of mass and momentum in the Boolean field, and may be considered as the microscopic mass-balance and momentum-balance equations, respectively, of the lattice gas.

2.7 Macrodynamical equations of the lattice gas

We now indicate how *macrodynamical* equations may be derived for the evolution of average quantities at a macroscopic scale for the lattice gas. To do so, we must define what we mean by "average" and "macroscopic," but it is instructive to first consider mass flux on the lattice in a manner analogous to our derivation of the equation of continuity (2.7).

Figure 2.6 is the same as Figure 1.1, but with an area $\mathcal{A}$ of lattice sites enclosed by a perimeter $\mathcal{S}$. From the conservation of mass given by equation (2.41) (or the evolution of the particles themselves—see the exercises), we may write

$$\sum_{\mathbf{x} \in \mathcal{A}} \sum_i [n_i(\mathbf{x}, t + 1) - n_i(\mathbf{x}, t)] = -(\text{net mass flux out of } \mathcal{S}). \tag{2.43}$$

The above expression is the microscopic analog of equation (2.4). We recognize the left-hand side as a kind of finite difference in time summed over the area $\mathcal{A}$, and the right-hand side as a verbal statement of a discrete representation of a surface integral over the perimeter $\mathcal{S}$. We seek an expression similar to equation (2.43) in terms of some average value of the n_i's.

We denote the average value of a quantity X by $\langle X \rangle$. In statistical physics, the definition of such an average is usually done in an axiomatic manner that assumes the existence of a probability distribution. This probability distribution is assumed to correspond to any of a number of measurable definitions. A common choice is to consider the average to be taken over an ensemble of experiments with different initial conditions. Rather than defining this average now, we simply assume that the average quantity $\langle n_i \rangle$ exists. That is not enough, however. We must also describe changes of scale so that we may consider the evolution of hydrodynamic quantities at temporal scales slow enough, and spatial scales large enough, such that $\langle n_i \rangle$ varies slowly in both space and time. Conceptually, the macroscopic time scale is much larger than one time step, and the macroscopic spatial scale is much larger than a lattice unit. In terms of our discussion of Figure 2.2, the macroscopic length scale is L_{hydro}.

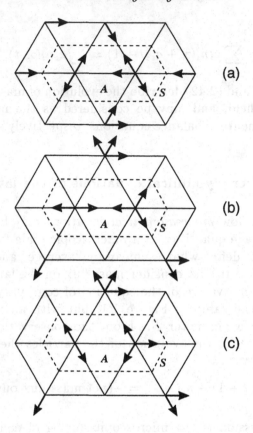

Fig. 2.6. The same as Figure 1.1, but with a region $\mathcal{A}$ of lattice sites enclosed by a perimeter $\mathcal{S}$. As before, the transition from (a) to (b) represents propagation, and the transition from (b) to (c) represents collisions.

Given such a description, we may identify $\sum_i \langle n_i \rangle$ with a slowly varying mass and $\sum_i \langle n_i \rangle \mathbf{c}_i$ with a slowly varying mass flux. The left-hand side of equation (2.43) may then be written as a time derivative of the mass and, following the same reasoning we used to derive equation (2.7), the right-hand side may be equated to the divergence of the mass flux. Specifically, we have

$$\partial_t \sum_i \langle n_i \rangle = -\partial_\alpha \sum_i \langle n_i \rangle c_{i\alpha}, \qquad (2.44)$$

where the α-component of the ith velocity vector $\mathbf{c}_i$ is given by $c_{i\alpha}$.

A similar argument may be given for each component of the momentum flux. From the conservation of momentum we may write

$$\sum_{\mathbf{x}\in\mathcal{A}}\sum_i [n_i(\mathbf{x},t+1) - n_i(\mathbf{x},t)]c_{i\alpha} = -(\text{net flux of } \alpha\text{-momentum out of } \mathcal{S}).$$

(2.45)

In terms of the slowly varying field $\langle n_i \rangle$, we recognize the left-hand side as the time derivative of the mass flux $\sum_i \langle n_i \rangle c_{i\alpha}$ and the right-hand side as the divergence of the momentum flux. Since the α-momentum carried by $\langle n_i \rangle$ in the β-direction is $\langle n_i \rangle c_{i\alpha}c_{i\beta}$, equation (2.45) may be rewritten for $\langle n_i \rangle$ as

$$\partial_t \sum_i \langle n_i \rangle c_{i\alpha} = -\partial_\beta \sum_i \langle n_i \rangle c_{i\alpha}c_{i\beta}.$$

(2.46)

To express equations (2.44) and (2.46) in terms of physical variables, we define the mass density

$$\rho = \sum_i \langle n_i \rangle,$$

(2.47)

and the momentum density

$$\rho u_\alpha = \sum_i \langle n_i \rangle c_{i\alpha}.$$

(2.48)

Substituting these definitions into (2.44) and (2.46), we then obtain the continuity equation,

$$\partial_t \rho = -\partial_\alpha(\rho u_\alpha),$$

(2.49)

and the macroscopic momentum-balance equation,

$$\partial_t(\rho u_\alpha) = -\partial_\beta \Pi^{(0)}_{\alpha\beta},$$

(2.50)

where in the latter case we have introduced the lattice analog of the inviscid momentum flux density tensor:

$$\Pi^{(0)}_{\alpha\beta} = \sum_i \langle n_i \rangle c_{i\alpha}c_{i\beta}.$$

(2.51)

We note that equation (2.50) is identical to the momentum-balance equation (2.18), but the momentum flux tensor (2.51) may not yet be identified as having the real-world form of equation (2.19). To obtain that correspondence we must proceed further.

Through symmetry arguments it is possible to find the general form of $\Pi^{(0)}_{\alpha\beta}$ when the lattice gas is flowing at low velocity. Indeed if the velocity

u is small, we may consider an expansion of $\Pi^{(0)}_{\alpha\beta}$ of the form[‡]

$$\Pi^{(0)}_{\alpha\beta} = p_0(\rho)\delta_{\alpha\beta} + \lambda_{\alpha\beta\gamma\delta}(\rho)u_\gamma u_\delta + \mathcal{O}(u^4). \tag{2.52}$$

In ordinary fluids the tensor $\lambda_{\alpha\beta\gamma\delta}$ is isotropic. However, because we work with an underlying lattice, this is not the case for lattice gases. In fact, $\lambda_{\alpha\beta\gamma\delta}$ acts instead as an elasticity tensor, and inherits the symmetry properties of the lattice just as elasticity tensors share the symmetry properties of a crystal lattice. This "memory" of the lattice would seem to make it impossible for the lattice gas to simulate a fluid, except for the remarkable property of hexagonal lattices that one may find in textbooks on elasticity theory. We quote Landau and Lifshitz: "It should be noticed that deformation in the xy-plane ... is determined by only two moduli of elasticity, as for an isotropic body; that is, the elastic properties of a hexagonal crystal are isotropic in the plane perpendicular to the [hexagonal] axis." In Appendix A, we discuss the general form of such an isotropic tensor. We find

$$\lambda_{\alpha\beta\gamma\delta} = A(\rho)\delta_{\alpha\beta}\delta_{\gamma\delta} + B'(\rho)(\delta_{\alpha\gamma}\delta_{\beta\delta} + \delta_{\alpha\delta}\delta_{\beta\gamma}), \tag{2.53}$$

where A and B' are two independent "elastic" moduli. In Chapter 3 we shall see that A and B' may be determined from the average density ρ. Once this is done, the momentum conservation equation takes the form

$$\partial_t(\rho u_\alpha) = -\partial_\alpha p' - \partial_\beta[B(\rho)\rho u_\alpha u_\beta] \tag{2.54}$$

where

$$B(\rho) = 2B'(\rho)/\rho, \qquad p' = p_0(\rho) + A(\rho)u^2. \tag{2.55}$$

Equation (2.54) is the lattice-gas analog of the real-world momentum-balance equation (2.17). We would immediately get more than an analogy if we were to find that $A = 0$ and $B = 1$. But that would be a lucky coincidence, and in general we have no such luck.

The problem of $B \neq 1$ deserves some further comment. The term in B is the divergence of the flux of momentum $\rho\mathbf{u}$ at velocity $B\mathbf{u}$, not $\mathbf{u}$. For incompressible flow, this corresponds to an advective derivative of the form $d\mathbf{u}/dt = \partial\mathbf{u}/\partial t + (B\mathbf{u} \cdot \nabla)\mathbf{u}$ instead of equation (2.11). This poses a problem that may be formally characterized as the breaking of *Galilean invariance*. Specifically, the form of equations (2.49) and (2.54) is not invariant under the the *Galilean transformation* $\mathbf{x} \to \mathbf{x} - \mathbf{w}t$, where **w** is the constant velocity of a moving frame of reference. Roughly speaking, the breaking of Galilean invariance may be traced to the fact that

---

[‡] In this expression, the terms which contain odd powers of u vanish, due to a symmetry property of the lattice which we discuss in Appendix A.

each particle may move with only one speed, thus making the stationary reference frame in some sense preferred over all others.

The lack of Galilean invariance is not quite as bad as one might think. When the velocity of a compressible fluid is low enough, the flow becomes nearly incompressible and equation (2.8) holds. Exactly how incompressible flow is approximated by low-velocity gas flow is discussed in Appendix D. One of the results of that discussion is a demonstration that the non-physical $\mathcal{O}(u^2)$ correction to the pressure in equation (2.54) is of no consequence, since, as we have already stated at the end of Section 2.3, the equation of state is irrelevant in incompressible flow. Moreover, as we have also already noted, we may consider the density ρ in equation (2.54) as constant. If we now perform the change of variables

$$\mathbf{v} = B(\rho)\mathbf{u}, \qquad p = B(\rho)p', \tag{2.56}$$

then substitution of equation (2.56) into equation (2.54) yields, keeping in mind that ρ is constant,

$$\partial_t \rho v_\alpha = -\partial_\alpha p - \partial_\beta(\rho v_\alpha v_\beta). \tag{2.57}$$

Equation (2.57) corresponds exactly to the real-world momentum-balance equation (2.17). It must be remembered, however, that the velocity $\mathbf{v}$ of equation (2.57) is not the velocity $\mathbf{u}$ defined by equation (2.48), but rather a rescaled velocity. One of the physical consequences of this distinction is considered in Exercise 2.4.

In the next chapter we will give a more precise derivation of the Euler equation of the lattice gas. In doing so, we will explicitly compute the coefficients A and B and the pressure p_0. Then in successive chapters we will quantify the viscous behavior of the lattice gas in addition to discussing other lattices that have the remarkable isotropy property (2.53) and that also lead to the correct incompressible equations.

2.8 Exercises

2.1 Consider the test volume V of Figure 2.3. Use the momentum-balance equation (2.18) to show that the momentum flux density tensor $\Pi_{\alpha\beta}^{(0)}$ gives the rate at which α-momentum flows out of a surface element $d\mathbf{S}$ of V when $d\mathbf{S}$ is aligned with the β-direction. Does this interpretation change if we use the viscous form of the momentum balance given by equation (2.29)?

2.2 Verify explicitly equations (2.43) and (2.45) for the specific configurations given in Figure 2.6.

2.3 Why were the details of the collision operator given by equations (2.37) and (2.35) not necessary for the the derivation of the hydrodynamic equations (2.49) and (2.50)? Would you expect the same to be true for the derivation of a viscous momentum-conservation equation?

2.4 Compute the curl of equation (2.54) and obtain an equation for the vorticity $\omega = \nabla \times \mathbf{u}$ in incompressible flow. Show that vorticity is advected at velocity $B\mathbf{u}$ rather than $\mathbf{u}$. Define a change of variables $\mathbf{v}(\mathbf{u})$ and $\omega'(\omega)$ such that your new vorticity equation matches the curl of equation (2.57).

2.9   Notes

Aspects and further details of our discussion of the continuum hypothesis and the derivation of the continuity, Euler, and Navier-Stokes equations may be found in the following textbooks:

2.1 Batchelor, G. K. (1967). *An Introduction to Fluid Dynamics* (Cambridge University Press, Cambridge).

2.2 Landau, L. D. and Lifshitz, E. M. (1959). *Fluid Mechanics* (Pergamon Press, New York).

2.3 Tritton, D. J. (1988). *Physical Fluid Dynamics, 2nd edition* (Clarendon Press, Oxford).

A nice elementary discussion of the molecular origin of viscosity is in:

2.4 Reif, F. (1965). *Fundamentals of Statistical and Thermal Physics* (McGraw-Hill, New York).

Readers desiring a review of tensors and their application to the equations of fluid dynamics may wish to consult:

2.5 Aris, R. (1962). *Vectors, Tensors, and the Basic Equations of Fluid Mechanics* (Prentice-Hall, Englewood Cliffs, N. J.).

The quotation from Landau and Lifshitz is on page 40 of:

2.6 Landau, L. D. and Lifshitz, E. M. (1959). *Theory of Elasticity* (Pergamon Press, New York).

3

Inviscid two-dimensional lattice-gas hydrodynamics

We now provide our first detailed derivation of the hydrodynamics of lattice gases. To keep matters from becoming unnecessarily complicated, we mostly restrict the discussion in this chapter to two-dimensional (2D) models. We begin with the simplest possible 2D model on a square lattice. We then repeat the calculation for the hexagonal lattice model. The principal result of this chapter is the derivation of the Euler equation of both models. This equation has the form already indicated in Chapter 2 but here the unknown coefficients are explicitly calculated. For future use we also include a general calculation in arbitrary dimension D and with an arbitrary number of rest particles.

The hydrodynamic behavior that we thus find at the macroscopic scale is a consequence of the existence of a kind of thermodynamic equilibrium. This equilibrium state is described by the Fermi-Dirac distribution of statistical mechanics. How this distribution arises is described in detail in Chapters 14 and 15. In this chapter we give a simpler derivation of some properties of equilibrium, which are sufficient to obtain the Euler equation.

3.1 Homogeneous equilibrium distribution on the square lattice

Our first task is to calculate $\langle n_i \rangle$, the average value of the Boolean variable n_i introduced in Section 2.6. Repeated applications of the rules of propagation and collision in the lattice gas cause these average particle populations to quickly reach an equilibrium state regardless of initial conditions. This state of equilibrium has several properties analogous to those of the more realistic classical models of gases. Here we will only need to know two such properties:

29

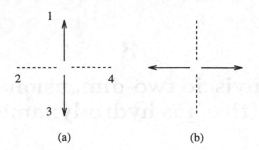

(a)   (b)

Fig. 3.1.   The collision operator for the HPP model exchanges configurations
(a) and (b) on the figure. For the collision to happen, the two directions not
represented in (a) and (b) must be empty. In equilibrium, there are as many
collisions going from (a) to (b) as there are going from (b) to (a).

- The distributions $\langle n_i \rangle$ are independent of spatial location. Thus we
  say that the equilibrium distribution is *homogeneous*.

- The distributions $\langle n_i \rangle$ depend only on the collisional invariants, that
  is, the total mass and momentum.

With these two properties we can compute the equilibrium distribution
of various models. For clarity, we begin with a particularly simple model,
the so-called HPP gas on a square lattice. (The initials HPP derive from
Hardy, de Pazzis, and Pomeau, the originators of the model). The particle
configurations that change due to collisions are illustrated in Figure 3.1.
The microdynamical equation that defines this model reads

$$n_i(\mathbf{x} + \mathbf{c}_i, t + 1) = n_i(\mathbf{x}, t) + \Delta_i[\mathbf{n}(\mathbf{x}, t)], \tag{3.1}$$

where the collision operator Δ_i is defined by

$$\Delta_i[\mathbf{n}] = \overline{n}_i \overline{n}_{i+2} n_{i-1} n_{i+1} - n_i n_{i+2} \overline{n}_{i-1} \overline{n}_{i+1}. \tag{3.2}$$

As in Chapter 2.6, we again use the notation $\overline{x} = 1 - x$ for any variable
x, and we now take the subscript i to imply "$i \bmod 4$" to account for the
four-fold symmetry of the square lattice. Taking the average of equation
(3.2) yields

$$\langle n_i n_{i+2} \overline{n}_{i-1} \overline{n}_{i+1} \rangle = \langle \overline{n}_i \overline{n}_{i+2} n_{i-1} n_{i+1} \rangle. \tag{3.3}$$

The term on the left hand side of equation (3.3) is the average number of
particles leaving the state in Figure 3.1a to enter the state in Figure 3.1b.
Thus equation (3.3) expresses the fact that fluxes from (a) to (b) and (b)
to (a) exactly balance in a state of equilibrium.

It is impossible to make progress on the basis of equation (3.3) alone
because there is no simple way to work with the averages of products.
However it is often possible to assume that this equilibrium is *factorized*.

A factorized equilibrium is one in which the averages of products can be expressed as products of averages. (Statistically, this statement means that the bits n_i are uncorrelated.) Then equation (3.3) reduces to

$$N_i N_{i+2} \overline{N}_{i-1} \overline{N}_{i+1} = \overline{N}_i \overline{N}_{i+2} N_{i-1} N_{i+1}, \qquad (3.4)$$

where we use the notation $N_i = \langle n_i \rangle$ to represent the mean particle densities.

As stated above, the equilibrium state is uniform and is parameterized by the average density ρ and momentum $\rho\mathbf{u}$. In terms of the mean particle densities N_i, we have

$$\sum_i N_i = \rho \qquad (3.5)$$

for the mass density, and

$$\sum_i N_i c_{ix} = \rho u_x \qquad (3.6)$$

$$\sum_i N_i c_{iy} = \rho u_y \qquad (3.7)$$

for the two components of momentum. Equations (3.4), (3.5), (3.6), and (3.7) are a closed system of four equations for the four mean populations N_i. They may therefore be solved exactly for the equilibrium state $N_i(\rho, \mathbf{u})$. Here, however, we adopt a pedagogical approach to set the stage for the more general developments that follow, and instead seek a solution by means of a simple series expansion.

3.2 Low-velocity equilibrium distribution

We assume that the modulus of the average velocity, $u = |\mathbf{u}|$, is small. We can then solve for N_i as a power series in u. The most general form that this series may have is

$$N_i = A_i + B_{i\alpha} u_\alpha + D_{i\alpha\beta} u_\alpha u_\beta + O(u^3). \qquad (3.8)$$

Here $i = 1, \ldots, 4$, A_i is an unknown scalar, $B_{i\alpha}$ is the α-component of an unknown vector $\mathbf{B}_i$, and $\mathbf{D}_i$ is an unknown second-rank tensor. Each of these unknowns must be determined from equations (3.4), (3.5), (3.6), and (3.7). However, such an analysis is cumbersome unless we simplify things using symmetry.

First, we notice that at zero velocity none of the four lattice directions is preferred. The state is as close to being isotropic as allowed by the lattice geometry. All distributions are therefore equal and from equation (3.5) we find

$$A_i = \rho/4, \qquad (3.9)$$

which also satisfies equation (3.4).

Second, we see that the vector $\mathbf{B}_i$ and the tensor $\mathbf{D}_i$ must have the same symmetry as the lattice, that is they must be unchanged in any symmetry operation that preserves the lattice and the direction i. For the square lattice this means the reflection that changes $i - 1$ into $i + 1$.

To fix ideas, consider the case $i = 1$, and assume that the direction $\mathbf{c}_1$ is parallel to the x-axis. The vector $\mathbf{B}_1$ must be invariant by the change of y into $-y$. It can thus only be aligned with $\mathbf{c}_1$. Therefore $\mathbf{B}_i = B\mathbf{c}_i$.

The symmetries of $\mathbf{D}_i$ are trickier. There is a general derivation of $\mathbf{D}_i$ in Section A.6. Here we give a simpler derivation in the special case of a square lattice. There are in principle $4 \times 2 \times 2 = 16$ coefficients. We can eliminate half of them by noting that the tensor $\mathbf{D}_1$ must also be invariant by the change of y into $-y$. As a result, any product $D_{1\alpha\beta}u_\alpha u_\beta$ that contains an odd number of terms with contributions in the y-direction must vanish. We are thus left with diagonal elements only:

$$\mathbf{D}_1 = \begin{pmatrix} D_{1xx} & 0 \\ 0 & D_{1yy} \end{pmatrix}. \tag{3.10}$$

The argument just given for $\mathbf{D}_1$ may be repeated for any $\mathbf{D}_i$, leaving us with eight coefficients. This number may be reduced even further, however. Since the second-order term in equation (3.8) must be invariant under the double reflection $x \to -x$ and $y \to -y$, we have

$$D_{1xx} = D_{3xx}, \qquad D_{2xx} = D_{4xx}, \qquad D_{1yy} = D_{3yy}, \qquad D_{2yy} = D_{4yy}. \tag{3.11}$$

Furthermore, there is also a symmetry under a $\pi/2$ rotation that gives

$$D_{1xx} = D_{2yy}, \qquad D_{1yy} = D_{2xx}. \tag{3.12}$$

Combining equations (3.11) and (3.12), we find six relations for the eight non-vanishing coefficients of $\mathbf{D}_i$. Each of these relations may be satisfied by $D_{i\alpha\beta} = \delta_{\alpha\beta}$. They are also consistent with $D_{i\alpha\beta} = c_{i\alpha}c_{i\beta}$. Since these two solutions are linearly independent, we write

$$D_{i\alpha\beta} = \lambda_1 c_{i\alpha}c_{i\beta} + \lambda_2 \delta_{\alpha\beta}, \tag{3.13}$$

where λ_1 and λ_2 are unknown coefficients.

The expansion (3.8) now reads

$$N_i = \rho/4 + Bc_{i\alpha}u_\alpha + (\lambda_1 c_{i\alpha}c_{i\beta} + \lambda_2 \delta_{\alpha\beta})u_\alpha u_\beta + \mathcal{O}(u^3). \tag{3.14}$$

Multiplying both sides by $c_{i\alpha}$, summing over i, and using the momentum equations (3.6) and (3.7), we find

$$\frac{\rho}{4}\sum_i c_{i\alpha} + B\sum_i c_{i\alpha}c_{i\beta}u_\beta + \lambda_1 \sum_i c_{i\alpha}c_{i\beta}c_{i\gamma}u_\beta u_\gamma + \lambda_2 \sum_i c_{i\alpha}\delta_{\beta\gamma}u_\beta u_\gamma = \rho u_\alpha, \tag{3.15}$$

where we have ignored terms of $\mathcal{O}(u^3)$. The first and last terms on the left-hand side vanish because $\sum_i c_{i\alpha} = 0$, an obvious relation that results from a lattice geometry in which for every $\mathbf{c}_i$ there is an opposite vector $\mathbf{c}_j = -\mathbf{c}_i$. The third term vanishes for the same reason. Thus we are left with

$$B \sum_i c_{i\alpha} c_{i\beta} u_\beta = \rho u_\alpha. \tag{3.16}$$

For the square lattice, one may easily show that $\sum_i c_{i\alpha} c_{i\beta} = 2\delta_{\alpha\beta}$. (The general result is given later in equation (A.33).) We thus find $B = \rho/2$.

Next, we insert the mass relation (3.5) into equation (3.14) and find

$$\rho + \sum_i (\lambda_1 c_{i\alpha} c_{i\beta} + \lambda_2 \delta_{\alpha\beta}) u_\alpha u_\beta = \rho \tag{3.17}$$

from which we obtain $\lambda_2 = -\lambda_1/2$. We then substitute our results for B and λ_2 into the expansion (3.14), and insert this long expression into the factorized equilibrium given by equation (3.4). After a laborious calculation, we obtain, neglecting terms above second order in u,

$$4\left(\frac{\rho}{4}\right)^2\left(1 - \frac{\rho}{2}\right)(u_x^2 - u_y^2) = \frac{\rho}{2}\left(1 - \frac{\rho}{4}\right)\lambda_1(u_x^2 - u_y^2), \tag{3.18}$$

from which we find

$$\lambda_1 = \frac{\rho}{2}\frac{1 - \rho/2}{1 - \rho/4}. \tag{3.19}$$

Finally we have the expansion

$$N_i = \frac{\rho}{4} + \frac{\rho}{2}\mathbf{u}\cdot\mathbf{c}_i + \frac{\rho}{2}\frac{1 - \rho/2}{1 - \rho/4}\left(c_{i\alpha} c_{i\beta} - \frac{1}{2}\delta_{\alpha\beta}\right)u_\alpha u_\beta + \mathcal{O}(u^3). \tag{3.20}$$

It is convenient for the purpose of simplifying this and other expressions in this book to introduce the density per link, or density per channel, $f = \rho/4$. Then

$$N_i = f\left[1 + 2\mathbf{u}\cdot\mathbf{c}_i + 2\frac{1 - 2f}{1 - f}\left(c_{i\alpha} c_{i\beta} - \frac{1}{2}\delta_{\alpha\beta}\right)u_\alpha u_\beta\right] + \mathcal{O}(u^3). \tag{3.21}$$

We are now ready to express momentum conservation for the lattice gas.

3.3 Euler equation on the square lattice

The preceding discussion was set in the context of the lattice gas equilibrium. However, when there is any sort of non-elementary flow (i.e., any motion other than uniform translation or rigid-body rotation) this assumption of equilibrium is no longer warranted. Nevertheless we may assume that the deviation from the equilibrium distributions N_i^{eq} provoked by the flow is small. More precisely, we make the following assumptions:

- The densities vary slowly in space and time, so $N_i(\mathbf{x})$ and $N_i(\mathbf{x}+\mathbf{c}_i)$ are nearly equal.

- The factorization assumption is still approximately valid.

These assumptions imply that the mass and momentum densities vary themselves slowly in space and time, for instance $u(\mathbf{x},t)$ and $u(\mathbf{x}+\mathbf{c}_i,t)$ are nearly equal. The second assumption allows the equilibrium distribution to approach the non-equilibrium N_i. With these assumptions the derivation of the distributions N_i can proceed as before. In other words, the equilibrium distribution remains a good approximation even when the lattice gas is not strictly in equilibrium.

This remark allows us to expand the inviscid momentum flux density tensor that we derived previously in Chapter 2.7. Substitution of the equilibrium distribution (3.21) for $\langle n_i \rangle$ in equation (2.51) gives the momentum flux to second order in velocity:

$$\Pi_{\alpha\beta} = 2f\delta_{\alpha\beta} + 2f\frac{1-2f}{1-f}\left(\sum_i c_{i\alpha}c_{i\beta}c_{i\gamma}c_{i\delta} - \delta_{\alpha\beta}\delta_{\gamma\delta}\right)u_\gamma u_\delta. \tag{3.22}$$

The tensor $\sum_i c_{i\alpha}c_{i\beta}c_{i\gamma}c_{i\delta}$ can be calculated easily for the square lattice after noticing that

$$\sum_i c_{i\alpha}c_{i\beta}c_{i\gamma}c_{i\delta} = 2 \tag{3.23}$$

when $\alpha = \beta = \gamma = \delta$, and

$$\sum_i c_{i\alpha}c_{i\beta}c_{i\gamma}c_{i\delta} = 0 \tag{3.24}$$

otherwise. Thus

$$\Pi_{xx} = 2f\left[1 + \frac{1-2f}{1-f}(2u_x^2 - u^2)\right], \tag{3.25}$$

$$\Pi_{xy} = 0, \tag{3.26}$$

and

$$\Pi_{yy} = 2f\left[1 + \frac{1-2f}{1-f}(2u_y^2 - u^2)\right], \tag{3.27}$$

where $u = |\mathbf{u}|$. Substitutions of these expressions into the macroscopic momentum-balance equation (2.50) then gives the governing equations for the hydrodynamic fields ρ and $\mathbf{u}$:

$$\partial_t(\rho u_x) + \partial_x(\Pi_{xx}) = 0, \qquad \partial_t(\rho u_y) + \partial_y(\Pi_{yy}) = 0. \qquad (3.28)$$

It is important to remark that these equations are not isotropic, which means that they are not invariant with any rotation of the plane, unlike the true Euler equation. Thus we now turn to a more general description of the lattice-gas equilibrium that will allow us to work with lattices that contain sufficient symmetry for isotropy.

3.4 Fermi-Dirac equilibrium distributions

The derivations we gave in Section 3.2 are relatively easy for the square lattice. For more general cases, however, we need another approach to the derivation of the probabilities N_i. Probability distributions in *statistical equilibrium*, or *equilibrium distributions*, are discussed in Chapter 14. There we show that the statistical equilibrium may be obtained provided we know the invariants of the lattice-gas model we consider. In this section, we will quickly derive the simplest form of the Fermi-Dirac statistical equilibrium using the mass and momentum invariants of the models.

First we need some definitions. The *local configuration* is the set of b bits describing the state of a lattice site. There are 4 bits for the square model and 6 for the hexagonal model. More complex models have multiple velocities, i. e. vectors $\mathbf{c}_i$ with several norms $|\mathbf{c}_i|$, or "live" in higher dimensions. We write $s(\mathbf{x}) = (s_i(\mathbf{x}))_i$ with $1 \leq i \leq b$ for the local configuration, on site $\mathbf{x}$, of the automaton. (We use the notation $(y_i)_i$ to indicate the entire vector, as opposed to a single component y_i. This is especially useful when multiply-indexed objects arise. We often use boldface for vectors, but not always: $(y_i)_i$ may be noted y or $\mathbf{y}$.). Then if there are $\mathcal{N}$ lattice points we define the *global configuration* of a lattice gas

$$s = (s(\mathbf{x}_1), \cdots, s(\mathbf{x}_\mathcal{N})). \qquad (3.29)$$

We also use the notation $\mathbf{n} = (n_i)_i$, as in Chapter 2, to describe a local configuration. A *global invariant* is a function J_s of the configuration s that stays unchanged in the course of the dynamics. We know already some *uniform global invariants*, such as the total mass and the total momentum on the lattice. These invariants are the sum of identical contributions for each site. They can be written in the general case of a D-dimensional space as

$$M_s = \sum_{\mathbf{x}} \sum_i s_i(\mathbf{x}) \quad \text{and} \quad \mathbf{G}_s = \sum_{\mathbf{x}} \sum_i s_i(\mathbf{x})\mathbf{c}_i \qquad (3.30)$$

where the sums over $\mathbf{x}$ are for all sites in the lattice $\mathcal{L}$. We may imagine models with additional invariants such as kinetic energy, chemical species, etc. It also happens that some lattice gases have additional invariants that complicate the analysis. We call them *spurious invariants* and discuss them in detail in Chapters 14 and 15. For the moment, however, we shall simplify the discussion assuming only mass and momentum invariants.

Statistical mechanics provides a general framework for the study of many-body systems with invariants. We give an overview of the foundations of statistical mechanics in Chapter 14. One of the most important results is the *Gibbs distribution*, which provides a statistical description of the microscopic fluctuations of a system in thermal equilibrium. Because the lattice gas exhibits a kind of thermodynamic equilibrium, its microscopic fluctuations may be described by the Gibbs probability distribution

$$P(s) = \frac{\exp\left(-hM_s - \mathbf{q} \cdot \mathbf{G}_s\right)}{\sum_\sigma \exp\left(-hM_\sigma - \mathbf{q} \cdot \mathbf{G}_\sigma\right)}. \tag{3.31}$$

Here the coefficients h and $\mathbf{q}$ are called the *chemical potentials* for mass and momentum. One sees immediately that the Gibbs distribution contains within it the statement of *equiprobability*, that is, that all configurations s with the same value of the conserved quantities have the same probability. A derivation of the Gibbs distribution is given in Appendix E.

We shall not need to evaluate the Gibbs distribution (3.31) directly. Instead, the Boolean nature of the lattice gas allows us to work with a simplification of the Gibbs distribution that applies to systems that satisfy an exclusion rule (i.e., systems composed of states that may be described only as being "occupied" or "unoccupied"). In this case, a consequence of the Gibbs distribution is the *Fermi-Dirac distribution*. For systems having only mass and momentum in their set of invariants, the Fermi-Dirac distribution reads

$$N_i^{eq} = \frac{1}{1 + \exp(h + \mathbf{q} \cdot \mathbf{c}_i)}, \tag{3.32}$$

where the superscript eq signifies the value of N_i in equilibrium. The derivation of (3.32) is done in Section 14.6. Expressions (3.5), (3.6), (3.7), and (3.32) define implicitly h and $\mathbf{q}$ as functions of ρ and $\mathbf{u}$. From $h(\rho, \mathbf{u})$ and $\mathbf{q}(\rho, \mathbf{u})$, one finds the equilibrium distribution $N_i^{eq}(\rho, \mathbf{u})$.

Usually we are unable to obtain a closed-form expression for $N_i^{eq}(\rho, \mathbf{u})$. As we did for the equilibrium distribution on the square lattice, however, we may evaluate it by means of a series expansion. We proceed by expanding h and $\mathbf{q}$ in powers of the velocity $\mathbf{u}$:

$$h(\rho, \mathbf{u}) = h_0 + h_2 u^2 + \mathcal{O}(u^4), \tag{3.33}$$

$$\mathbf{q}(\rho, \mathbf{u}) = q_1\mathbf{u} + \mathcal{O}(u^3). \tag{3.34}$$

To obtain these expressions, we used invariance of the populations under the change $\mathbf{u} \to -\mathbf{u}$, $\mathbf{c}_i \to -\mathbf{c}_i$, which implies that h must be an even function of $\mathbf{u}$ and $\mathbf{q}$ must be odd.

The Fermi-Dirac distribution (3.32) may be written $N_i^{eq} = \mathcal{F}(h + \mathbf{q} \cdot \mathbf{c}_i)$ where

$$\mathcal{F}(x) = \frac{1}{1 + e^x}. \tag{3.35}$$

A Taylor expansion around $x = h_0$ then gives

$$N_i^{eq} = \mathcal{F}(h_0) + \mathcal{F}'(h_0)(q_1\mathbf{u}\cdot\mathbf{c}_i + h_2 u^2) + \frac{1}{2}\mathcal{F}''(h_0)q_1^2(\mathbf{u}\cdot\mathbf{c}_i)^2 + \mathcal{O}(u^3). \tag{3.36}$$

Summing this expression over i for the case $\mathbf{u} = 0$, we immediately obtain $\mathcal{F}(h_0) = \rho/b$ from the mass relation (3.5). Using the notation $f = \rho/b$ for the "reduced" density and noting that $\mathcal{F}' = -\mathcal{F}(1 - \mathcal{F})$, we get

$$N_i^{eq} = f\left[1 - (1 - f)\left(q_1\mathbf{u}\cdot\mathbf{c}_i + h_2 u^2\right) + \frac{1}{2}(1 - f)(1 - 2f)q_1^2(\mathbf{u}\cdot\mathbf{c}_i)^2\right], \tag{3.37}$$

where we have neglected terms of $\mathcal{O}(u^3)$. Replacing the above expansion into the mass relation (3.5) and the momentum equations (3.6) and (3.7), we obtain two relations,

$$0 = -f\overline{f}q_1 u_\alpha C_\alpha^{(1)} + \left(\frac{1}{2}f\overline{f}(1 - 2f)q_1^2 C_{\alpha\beta}^{(2)} - bf\overline{f}h_2\delta_{\alpha\beta}\right)u_\alpha u_\beta, \tag{3.38}$$

and

$$\rho u_\alpha = fC_\alpha^{(1)} - f\overline{f}q_1 u_\beta C_{\alpha\beta}^{(2)}$$
$$+ \left(\frac{1}{2}f\overline{f}(1 - 2f)q_1^2 C_{\alpha\beta\gamma}^{(3)} - bf\overline{f}h_2 C_\alpha^{(1)}\delta_{\beta\gamma}\right)u_\beta u_\gamma, \tag{3.39}$$

where we have implicitly defined the tensors

$$C_{\alpha_1\dots\alpha_r}^{(r)} = \sum_i c_{i\alpha_1}\dots c_{i\alpha_r}. \tag{3.40}$$

At this point it is convenient to give the general D-dimensional result. From Appendix A we have the result

$$C_{\alpha\beta}^{(2)} = \frac{bc^2}{D}\delta_{\alpha\beta}, \tag{3.41}$$

where $c = |\mathbf{c}_i|$. Furthermore all odd-order r-tensors of the form $\mathbf{C}^{(r)}$ vanish.

Using these results, q_1 and h_2 may be found. The final result is

$$N_i^{eq} = f\left[1 + \frac{D}{c^2}c_{i\alpha}u_\alpha + G(f)Q_{i\alpha\beta}u_\alpha u_\beta\right] + \mathcal{O}(u^3), \tag{3.42}$$

where $i = 1$ to b, and

$$G(f) = \frac{D^2}{2c^4}\left(\frac{1 - 2f}{1 - f}\right), \quad \text{and} \quad Q_{i\alpha\beta} = c_{i\alpha}c_{i\beta} - \frac{c^2}{D}\delta_{\alpha\beta}. \quad (3.43)$$

It is interesting to remark that the coefficient $G(f)$ is symmetric by the change $f \to 1 - f$. This symmetry is a kind of particle-hole symmetry. However, the original model need not have this symmetry. For instance, the FHP model introduced in Chapter 2 is not symmetric if all bits are reversed. It is also worth noticing that for each α and β, the b-vector $(Q_{i\alpha\beta})_i$ is orthogonal to the mass and momentum b-vectors. Lastly, comparison of equation (3.42) with equation (3.21) shows that our pedagogical derivation of the equilibrium distribution on the square lattice gave a special case of the more general Fermi-Dirac equilibrium.

3.5   Euler equation for the hexagonal lattice

The Fermi-Dirac equilibrium distribution (3.42) may be used to obtain the Euler equation for the hexagonal lattice. Proceeding in a manner analogous to our calculations for the square lattice in Chapter 3.3, we see that the inviscid momentum flux density tensor now reads

$$\Pi^{(0)}_{\alpha\beta} = \sum_i N_i^{eq} c_{i\alpha}c_{i\beta}. \quad (3.44)$$

After substituting equation (3.42) for N_i^{eq}, the momentum-balance equation (2.50) then gives, for $b = 6$, $D = 2$, and $c = 1$,

$$\partial_t \rho u_\alpha + \partial_\beta \left[f C^{(2)}_{\alpha\beta} + f G(f)\left(C^{(4)}_{\alpha\beta\gamma\delta} - \frac{1}{2}\delta_{\gamma\delta}C^{(2)}_{\alpha\beta}\right) u_\gamma u_\delta \right] = \mathcal{O}(u^4). \quad (3.45)$$

The right-hand side contains terms of $\mathcal{O}(u^4)$ rather than $\mathcal{O}(u^3)$ because of invariance under the reflections $\mathbf{u} \to -\mathbf{u}$, $\mathbf{c}_i \to -\mathbf{c}_i$. The simplest approach to the evaluation of the tensor $\mathbf{C}^{(4)}$ in the above expression is to compute it explicitly for the case of the hexagonal lattice. We obtain the isotropic form

$$C^{(4)}_{\alpha\beta\gamma\delta} = \frac{3}{4}(\delta_{\alpha\beta}\delta_{\gamma\delta} + \delta_{\alpha\delta}\delta_{\beta\gamma} + \delta_{\alpha\gamma}\delta_{\beta\delta}), \quad (3.46)$$

which we then insert into equation (3.45) to obtain

$$\partial_t \rho u_\alpha + \partial_\beta[g(\rho)\rho u_\alpha u_\beta] = -\partial_\alpha[p(\rho, u^2)] + \mathcal{O}(u^4). \quad (3.47)$$

The parameter $g(\rho)$ and the pressure $p(\rho, u^2)$ are generally expressed in terms of the coefficients in the low velocity expansion (3.42). For the hexagonal lattice these parameters are

$$g(\rho) = \frac{3 - \rho}{6 - \rho}, \quad (3.48)$$

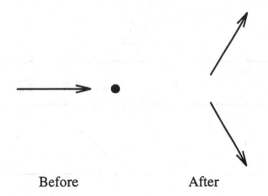

Before   After

Fig. 3.2. A collision involving a rest particle. Notice that momentum and mass are conserved.

$$p(\rho, u^2) = \frac{\rho}{2}\left[1 - g(\rho)u^2\right].\tag{3.49}$$

The Euler equation (3.47) together with the pressure-density relation (3.49) are somewhat different from the real-world equations derived in Chapter 2. However, as shown in Appendix D the correct incompressible equations are recovered at low Mach number. This is performed using the change of variable $\mathbf{u'} = g(\rho)\mathbf{u}$ already indicated in Section 2.7.

3.6 The 7-velocity model

A simple and very useful modification of the 6-velocity model of Chapter 2 may be obtained by adding a seventh particle with zero velocity. While ordinary particles hop to neighboring sites in the direction of their velocity, rest particles stay in place during the propagation step. During the collision step, rest particles may be involved in a number of new collisions (Figure 3.2). To list all the possible collisions in the 7-velocity model, the idea of *local mass-momentum packets* is useful. Figures 3.3 and 3.4 list all the packets of the 7-velocity model. The packets may be noted (n, g_x, g_y) where n is the number of particles, equal to the mass, and $\mathbf{g}$ is the momentum. In this figure, we describe the configurations only up to a rotation or reflection. We have distinguished two subgroups of packet (3,0,0), which we call $(3, 0, 0)^A$ and $(3, 0, 0)^B$.

The *collision-saturated 7-velocity model* (also known as FHP-III) has the following collision rules. Configurations are transformed into any of the *other* configurations of the same packet (n, g_x, g_y). However in some cases, such as packet $(3, 0, 0)$, the collision output is chosen to be another member of the same subpacket, either $(3, 0, 0)^A$ or $(3, 0, 0)^B$. There are at most $p = 3$ members of a subpacket in this scheme. Thus there are at

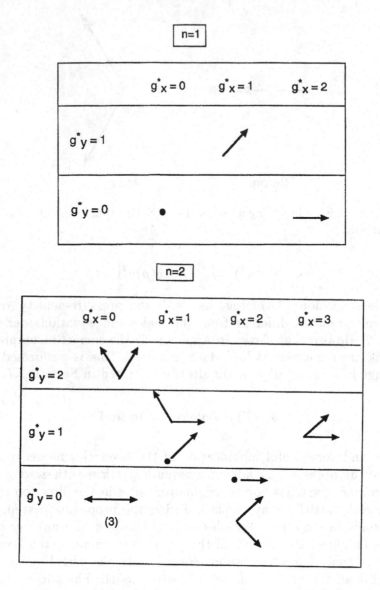

Fig. 3.3. A table of configurations with one or two particles for the 7-velocity FHP lattice gas [1.19]. The dot represents a rest particle.

most 2 outputs to choose from and the choice is achieved with a single random bit. The coding may be performed using table-lookup or Boolean operators. (These codes may be retrieved by the means outlined in the preface).

Another model is the *random-collision 7-velocity model*. A configuration in a packet is transformed into a configuration chosen at random in

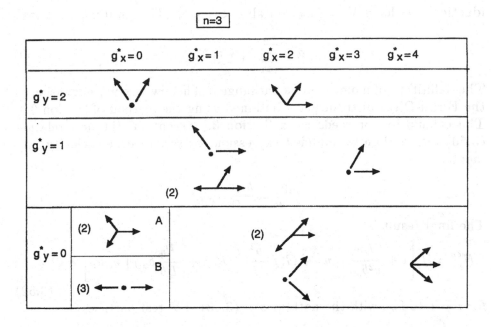

Fig. 3.4. Configurations with 3 particles for the 7-velocity FHP lattice gas. Notice that for $(n, g_x, g_y) = (3, 0, 0)$ there are two subclasses A and B [1.19]. In each subclass the configurations may be deduced from each other by rotations and reflections.

the same packet. One may choose the original configuration. There are at most $p = 5$ configurations to choose from (Figure 3.4). In practice a random number generator is used to perform the choice.

3.7 Euler equation with rest particles

The bit associated with the rest particle is noted n_0 and its velocity is $c_0 = 0$ while $c_1, \ldots, c_6$ are as before the velocities of moving particles. In some cases it is interesting to generalize the 7-velocity model to include an arbitrary number of rest particles. Consider models with b_r rest particles and $b_m = b - b_r$ moving particles. We then write the bits associated with rest particles n_{0k} with $0 < k \le b_r$. The state of a site is $s = (n_{01}, \ldots, n_{0b_r}, n_1, \ldots, n_{b_m})$. We assume here that the rest particles are identical and distinguishable, so each variable n_{0k} is of a Fermionic nature and the Fermi-Dirac distribution (3.32) still holds. Since rest particles are identical, we have $N_0 = \langle n_{0k} \rangle$ for all $0 < k \le b_r$. The total mass is then

$$\rho = \sum_{i=1}^{b_m} N_i + b_r N_0. \qquad (3.50)$$

The definition of momentum is unchanged. The low velocity expansion of the Fermi-Dirac distribution is obtained using the method of Section 3.4. The calculation proceeds as in Section 3.4 except for the new relation (3.50) and a slightly modified expression for second order velocity moments,

$$C_{\alpha\beta}^{(2)} = \frac{b_m c^2}{D} \delta_{\alpha\beta}. \qquad (3.51)$$

The final result is

$$N_i^{eq} = f \left[1 + \frac{Db}{c^2 b_m} c_{i\alpha} u_\alpha + G(f) \frac{b^2}{b_m^2} \left(Q_{i\alpha\beta} + \frac{c^2 b_r}{Db} \delta_{\alpha\beta} \right) u_\alpha u_\beta \right] + \mathcal{O}(u^3) \qquad (3.52)$$

for $i = 1$ to b_m, with the notations of (3.43). For rest particles

$$N_0^{eq} = f \left[1 - G(f) \frac{bc^2}{Db_m} u^2 \right] + \mathcal{O}(u^2) \qquad (3.53)$$

where $G(f)$ and $Q_{i\alpha\beta}$ are still given by (3.43).

The Euler equation is obtained by replacing (3.52) and (3.53) into the momentum-balance equation once again. For more generality we shall consider an arbitrary model dimension D. The calculation involves the tensor $\mathbf{C}^{(4)}$ which is evaluated in Appendix A for a general lattice having isotropic fourth order tensors (Property 5.4). The result generalizes our previous result (3.46) and is

$$C_{\alpha\beta\gamma\delta}^{(4)} = \frac{b_m c^4}{D(D+2)} (\delta_{\alpha\beta}\delta_{\gamma\delta} + \delta_{\alpha\delta}\delta_{\beta\gamma} + \delta_{\alpha\gamma}\delta_{\beta\delta}). \qquad (3.54)$$

After a short calculation we obtain

$$\partial_t \rho u_\alpha + \partial_\beta [g(\rho)\rho u_\alpha u_\beta] = -\partial_\alpha [p(\rho, u^2)] + \mathcal{O}(u^4). \qquad (3.55)$$

The parameters are

$$g(\rho) = \frac{bD}{b_m(D+2)} \frac{1 - 2\rho/b}{1 - \rho/b} \qquad (3.56)$$

$$p(\rho, u^2) = c_s^2 \rho - \rho g(\rho) \frac{c_s^2}{c^2} \left(1 + \frac{D}{2} - \frac{c^2}{2c_s^2} \right) u^2 \qquad (3.57)$$

where $c_s^2 = b_m c^2 / (bD)$. The quantity c_s is in fact the soundspeed, since $c_s^2 = \partial p / \partial \rho$. (Sound is discussed in more detail in Appendix D.) This equation has to be coupled as usual with the mass-conservation equation

$$\partial_t \rho + \nabla \cdot \rho \mathbf{u} = 0. \tag{3.58}$$

Why add rest particles? One advantage is the higher collision rate. It yields a faster relaxation to statistical equilibrium and a lower viscosity as discussed in Chapter 4. There is another advantage to the addition of rest particles, which appears now that we have computed the $g(\rho)$ factor. With large numbers of rest particles b/b_m in equation (3.56) may be large enough to find a density $\rho_0 > 0$ such that $g(\rho_0) = 1$. Computing around this density allows for more realistic Boolean calculations. This advantage may be important for flows with several species of particles, which are introduced later in this book. For flows with a single species, there is still the change of variable described in Section 2.7 and Appendix D that yields the Euler equations with $g = 1$.

3.8   Exercises

3.1 Check the calculation of λ_1 in equation (3.19). (This is rather lengthy, the main point of the exercise being that it generates appreciation for the wonderful simplification brought by the Fermi-Dirac distribution).

3.2 Check the calculation of $C^{(4)}$ in (3.46), without using the derivation in Appendix A. One possible shortcut is to represent the $\mathbf{c}_i$ as powers of the complex number $z = 1/2 + i\sqrt{3}/2$.

3.3 Consider one time step of the motion of a given particle in the 7-velocity model. Compute the probability that it collides with just one other particle.

3.4 Compute the probability of triple collisions in the 7-velocity model.

3.5 Check the derivation of expressions (3.52) and (3.53), then check expressions (3.55), (3.56) and (3.57).

3.9   Notes

The original references for the square lattice model are:

3.1 Hardy, J., Pomeau, Y., and de Pazzis, O. (1973). Time evolution of a two-dimensional model system. I. Invariant states and time correlation functions. *J. Math. Phys.* **14**, 1746–1759.

3.2 Hardy, J., de Pazzis, O., and Pomeau, Y. (1976). Molecular dynamics of a classical lattice gas: Transport properties and time correlation functions. *Phys. Rev. A.* **13**, 1949–1961.

The first exhaustive description of models with rest particles was performed by:

3.3 d'Humières, D. and Lallemand, P. (1987). Numerical simulations of hydrodynamics with lattice gas automata in two dimensions. *Complex Systems* **1**, 599–632.

Numerical experience confirms the theoretical arguments and shows that having at least one rest particle helps improve the physical behavior of the models considerably. A model with 18 rest particles was introduced, for the purpose of restoring Galilean invariance, by:

3.4 Gunstensen, A. K. and Rothman, D. H. (1991). A Galilean-invariant two-phase lattice gas. *Physica D* **47**, 53–63.

Models with rest particles are a special case of models with multiple velocity moduli $c_i = |\mathbf{c}_i|$. Such models allow for the conservation of kinetic energy in collisions

$$K = \frac{1}{2} \sum_i \mathbf{c}_i^2. \tag{3.59}$$

The Fermi-Dirac distribution is then modified to read

$$N_i^{eq} = \frac{1}{1 + \exp(h + \mathbf{q} \cdot \mathbf{c}_i + \beta \mathbf{c}_i^2 / 2)}. \tag{3.60}$$

with the usual notations, except for β which is the inverse temperature $\beta = 1/(k_B T)$ with k_B Boltzmann's constant. The Euler equations for such models are much harder to derive, although the derivation follows the general pattern in this chapter. An example of such a model may be found in:

3.5 Grosfils, P., Boon, J. P., Brito, R., and Ernst, M. H. (1993). Statistical hydrodynamics of lattice gas automata. *Phys. Rev. E* **48**, 2655–2668.

In such multiple-velocity lattice-gases, several other spurious factors may appear beyond $g(\rho)$. Thus there are violations of thermodynamics or Galilean invariance. For instance the quantity $e + p/\rho + \frac{1}{2}u^2$ should be constant along streamlines in inviscid steady flow (where $e = \sum_i N_i \mathbf{c}_i^2/(2\rho)$ is the specific kinetic energy, as in Appendix C). This property is proven for instance in:

3.6 Landau, L. D. and Lifshitz, E. M. (1959). *Fluid Mechanics* (Pergamon Press, New York).

This property will be violated by the Euler equations of the multiple-velocity lattice-gas model because of the spurious factors. The origins of what amounts to a violation of thermodynamics are explained using a simple piston thought experiment in:

3.7 Cercignani, C. (1994). On the thermodynamics of a discrete velocity gas, *Transport Theory and Statistical Physics* - Proceedings of Euromech 287 - Discrete Models in Fluid Dynamics **23**, 1-8.

Thus the construction of a model that correctly reproduces the compressible flow equations of Appendix C remains a challenge.

4

Viscous two-dimensional hydrodynamics

We now show how to derive the viscous hydrodynamics of the lattice gas, and to calculate an approximation to its viscosity. The viscosity is calculated within the *Boltzmann approximation* for the lattice gas. This approximation is an extension to time-dependent states of the factorization assumption of Section 3.1. As in Chapter 3, we first perform a pedagogical calculation for the square lattice. The analogous calculation for the hexagonal lattice is then performed. The general derivation of the Navier-Stokes equation and the calculation of viscosity for an arbitrary linear operator in the Boltzmann approximation is given later in Chapter 15.

4.1 Navier-Stokes equation on the square lattice

As we discussed in Section 2.5, in real Newtonian fluids the viscous stress is linearly related to the rate of deformation. We may postulate the same general relation for lattice gases. Thus we again assume that

$$\Pi^{\text{visc}}_{\alpha\beta} = -\tilde{\mu}_{\alpha\beta\gamma\delta} E_{\gamma\delta}, \tag{4.1}$$

where $E_{\alpha\beta} = \partial_\alpha u_\beta + \partial_\beta u_\alpha$ is again the rate-of-deformation tensor, $\Pi^{\text{visc}}_{\alpha\beta}$ is the viscous stress tensor, and $\tilde{\mu}_{\alpha\beta\gamma\delta}$ is a fourth-rank tensor that contains within it the "constitutive properties" of the lattice gas—specifically, symmetry properties and viscosity coefficients.

Once we have an expression for $\Pi^{\text{visc}}_{\alpha\beta}$, we may write the Navier-Stokes equation of the lattice gas, i.e., the viscous form of the momentum-balance equation (3.47) that we derived in the previous section. As we showed in Chapter 2.5, the general form of this equation is

$$\partial_t(\rho u_\alpha) = -\partial_\beta \left(\Pi^{(0)}_{\alpha\beta} + \Pi^{\text{visc}}_{\alpha\beta} \right). \tag{4.2}$$

The inviscid momentum flux density tensor $\Pi^{(0)}_{\alpha\beta}$ was defined in equation

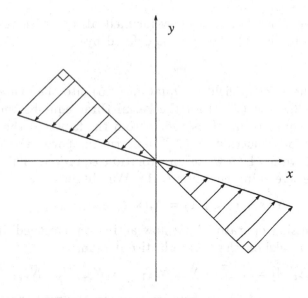

Fig. 4.1. A parallel shear flow, inclined at 45 degrees with the horizontal axis, may be used to investigate viscous effects in the HPP model.

(3.44) in terms of the equilibrium populations N_i^{eq}. The viscous stress tensor may be written in the analogous form

$$\Pi_{\alpha\beta}^{\text{visc}} = \sum_i N_i^{neq} c_{i\alpha} c_{i\beta}, \qquad (4.3)$$

where $N_i^{neq} = N_i - N_i^{eq}$ is a small correction to the equilibrium distribution due to a non-uniform flow. The stress induced by this small perturbation is proportional to the viscosity.

May we assume that the viscosity is isotropic, and not dependent on the orientation of a shear flow? If the lattice is sufficiently symmetric, then $\tilde{\mu}_{\alpha\beta\gamma\delta}$ will be itself isotropic, and the viscous stress tensor will have the form

$$\Pi_{\alpha\beta}^{\text{visc}} = -\mu_1 \left(\partial_\alpha u_\beta + \partial_\beta u_\alpha \right) - \mu_2 \partial_\gamma u_\gamma \delta_{\alpha\beta}, \qquad (4.4)$$

just as we derived for real fluids in Chapter 2.5. The question of "sufficient symmetry" turns out to be quite similar to our derivations of the Euler equation on the square and hexagonal lattices. In other words, $\tilde{\mu}_{\alpha\beta\gamma\delta}$ is anisotropic for the square lattice, but isotropic for the hexagonal lattice.

To find explicitly the tensor $\Pi_{\alpha\beta}^{\text{visc}}$ and the viscosity coefficients μ_i, a powerful theoretical tool, known as the Chapman-Enskog expansion, may be employed. In the following developments, however, we will simply calculate the shear viscosity by computing the distribution N_i^{neq}. Below, we consider a specific case on the square lattice.

We consider a parallel shear flow inclined at $\pi/4$ with respect to the lattice axes (Figure 4.1). The flow is defined by

$$u_x = u_y = \omega(x - y). \tag{4.5}$$

If we insert the velocity field of equation (4.5) into the expression found in equations (3.21) or (3.42) for the Fermi-Dirac equilibrium, we do not obtain a new equilibrium. This comes from the fact that the equilibrium was obtained for a distribution N_i^{eq} uniform in space. For non-uniform distributions, we need to write an evolution equation. This equation is obtained after averaging equation (3.1). We obtain

$$N_i(\mathbf{x} + \mathbf{c}_i, t + 1) = N_i(\mathbf{x}, t) + \Delta_i[\mathbf{N}(\mathbf{x}, t)] \tag{4.6}$$

where the collision operator Δ_i is now acting on averaged distributions. For the HPP model on the square lattice it reads

$$\Delta_i[\mathbf{N}] = \overline{N}_i \overline{N}_{i+2} N_{i-1} N_{i+1} - N_i N_{i+2} \overline{N}_{i-1} \overline{N}_{i+1}. \tag{4.7}$$

In writing equation (4.7), we have once again assumed that the average populations N_i are uncorrelated. Such an assumption is known as the *molecular chaos approximation*, and equations such as (4.6) that employ it are known as *Boltzmann equations*. We now proceed to solve equation (4.6) approximately by linearizing the nonlinear collision operator (4.7).

If we insert the expression (4.5) in to the Fermi-Dirac expansion we do not have an exact solution of equation (4.6), but an approximate one when ω is small. It is then natural to make an expansion for low shear rates and write

$$N_i(\mathbf{x}) = N_i^{eq}(\mathbf{x}; \mathbf{u}) + \epsilon_i + \mathcal{O}(\omega^2) \tag{4.8}$$

where $N_i^{eq}(\mathbf{x}; \mathbf{u})$ is obtained from the equilibrium distribution (3.21) and ϵ_i is an unknown correction of order ω. We insert this expansion into the collision operator of equation (4.6). The steady-state, time-independent form of the Boltzmann equation is then

$$N_i(\mathbf{x} + \mathbf{c}_i) - N_i(\mathbf{x}) = \Delta_i[\mathbf{N}^{eq}(\mathbf{x}; \mathbf{u})] + \sum_j \frac{\partial \Delta_i}{\partial N_j}\bigg|^{eq} \epsilon_j + \mathcal{O}(\omega^2) \tag{4.9}$$

where the *eq* superscript means that the derivative is taken for $\mathbf{N} = \mathbf{N}^{eq}(\mathbf{x}; \mathbf{u})$. In this expression the first term on the right-hand side vanishes because it just expresses the local collision operator in the equilibrium case. The second term is the derivative of the collision operator estimated near the equilibrium at velocity $\mathbf{u}$. At first order we can approximate it with the derivative estimated near zero velocity. We shall use the notation

$$\Lambda_{ij} = \frac{\partial \Delta_i}{\partial N_j}\bigg|_{\mathbf{N}=\mathbf{N}^{eq}(\mathbf{x};0)} \tag{4.10}$$

After inserting the expansion (4.8) into the left-hand side of equation (4.9), we obtain

$$N_i^{eq}(\mathbf{x} + \mathbf{c}_i) - N_i^{eq}(\mathbf{x}) = \sum_j \Lambda_{ij}\epsilon_j + \mathcal{O}(\omega^2). \qquad (4.11)$$

We now estimate N_i^{eq} from the equilibrium distribution (3.21). We find

$$N_1^{eq} = f[1 + 2\omega(x - y)] + \mathcal{O}(\omega^2) \qquad (4.12)$$
$$N_2^{eq} = f[1 + 2\omega(x - y)] + \mathcal{O}(\omega^2) \qquad (4.13)$$
$$N_3^{eq} = f[1 - 2\omega(x - y)] + \mathcal{O}(\omega^2) \qquad (4.14)$$
$$N_4^{eq} = f[1 - 2\omega(x - y)] + \mathcal{O}(\omega^2) \qquad (4.15)$$

Substitution of these expressions into equation (4.11) shows that $N_i^{eq}(\mathbf{x} + \mathbf{c}_i) - N_i^{eq}(\mathbf{x})$ is of order ω. This may be checked once we solve for the ϵ_i. Thus, after dropping second order terms from equation (4.11), we have the linear expression

$$\begin{pmatrix} 2f\omega \\ -2f\omega \\ 2f\omega \\ -2f\omega \end{pmatrix} = \Lambda \begin{pmatrix} \epsilon_1 \\ \epsilon_2 \\ \epsilon_3 \\ \epsilon_4 \end{pmatrix}. \qquad (4.16)$$

Before we solve equation (4.16), we need as a first step to analyze the structure of the linearized operator Λ. From equation (4.7), the linearized collision operator about a general distribution N_i reads

$$\Lambda_{11} = -\overline{N}_3 N_2 N_4 - N_3 \overline{N}_2 \overline{N}_4 \qquad (4.17)$$
$$\Lambda_{12} = \overline{N}_1 \overline{N}_3 N_4 + N_1 N_3 \overline{N}_4 \qquad (4.18)$$
$$\Lambda_{13} = -\overline{N}_1 N_2 N_4 - N_1 \overline{N}_2 \overline{N}_4 \qquad (4.19)$$
$$\Lambda_{14} = \overline{N}_1 \overline{N}_3 N_2 + N_1 N_3 \overline{N}_2 \qquad (4.20)$$

The other lines are obtained by using the $\pi/2$ symmetry of the lattice and of the collision operator. Because of reflection symmetries about the $\mathbf{c}_i$ vectors we also find that the matrix is symmetric. We thus find that Λ has the form of a symmetric circulant matrix, i.e.,

$$\Lambda_{ij} = L_{|i-j|} \qquad (4.21)$$

where the L_k are a vector of coefficients. If we specialize to the zero velocity equilibrium distribution $N_i^{eq} = f$ we get

$$\Lambda = f\overline{f} \begin{pmatrix} -1 & 1 & -1 & 1 \\ 1 & -1 & 1 & -1 \\ -1 & 1 & -1 & 1 \\ 1 & -1 & 1 & -1 \end{pmatrix}. \qquad (4.22)$$

Because this operator is symmetric, it has only real eigenvalues. There are four distinct eigenvectors:

$$\mathbf{1} = \begin{pmatrix} 1 \\ 1 \\ 1 \\ 1 \end{pmatrix}, \qquad (c_{ix})_i = \begin{pmatrix} 1 \\ 0 \\ -1 \\ 0 \end{pmatrix} \tag{4.23}$$

$$(c_{iy})_i = \begin{pmatrix} 0 \\ 1 \\ 0 \\ -1 \end{pmatrix}, \qquad \mathbf{v} = \begin{pmatrix} 1 \\ -1 \\ 1 \\ -1 \end{pmatrix}. \tag{4.24}$$

The first three eigenvectors have zero eigenvalues, and $\mathbf{v}$ has a non-zero eigenvalue $\lambda = -4f\overline{f}$. (The $\mathbf{v}$ eigenvector is also seen to be related to the c_i by $v_i = 2Q_{ixx} = -2Q_{iyy}$.) Since $\mathbf{v}$ is a multiple of the left-hand side of equation (4.16), we find

$$2f\omega v_i = \sum_j \Lambda_{ij}\epsilon_j = -4f\overline{f}\epsilon_i \tag{4.25}$$

or

$$\begin{pmatrix} \epsilon_1 \\ \epsilon_2 \\ \epsilon_3 \\ \epsilon_4 \end{pmatrix} = -\frac{\omega}{2\overline{f}} \begin{pmatrix} 1 \\ -1 \\ 1 \\ -1 \end{pmatrix}. \tag{4.26}$$

The correction ϵ_i to the equilibrium distribution N_i^{eq} may be used to estimate the new momentum flux. We identify ϵ_i as the first-order approximation of $N_i^{neq} = N_i - N_i^{eq}$. Substitution into equation (4.3) then gives an explicit expression for the viscous stress tensor:

$$\Pi_{\alpha\beta}^{\text{visc}} = \sum_i \epsilon_i c_{i\alpha} c_{i\beta}. \tag{4.27}$$

Thus we find

$$\Pi_{xx}^{\text{visc}} = -\frac{\omega}{1-f} \tag{4.28}$$

$$\Pi_{yy}^{\text{visc}} = \frac{\omega}{1-f}. \tag{4.29}$$

To find the viscosity coefficient, we note that in a shear flow of the form (4.5), we expect from equation (4.4) that

$$\Pi_{xx}^{\text{visc}} = -2\mu_1\omega \tag{4.30}$$

$$\Pi_{yy}^{\text{visc}} = 2\mu_1\omega. \tag{4.31}$$

Thus we find the viscosity coefficient $\mu_1 = 1/(2\overline{f})$. The kinematic viscosity, defined earlier in equation (2.31), is

$$\nu = -\frac{1}{2\lambda}, \tag{4.32}$$

where we have made the substitution $\lambda = -4f\overline{f}$. (There is a correction to this result which we discuss later in in Section 4.3 in the context of the hexagonal model.)

We close our discussion of the viscous equations on the square lattice with a word of warning. The HPP gas is a very peculiar type of lattice gas. We have already seen that the HPP model does not yield isotropic fourth order tensors. Moreover, in addition to mass and momentum the HPP model has many spurious invariants, as shown in Chapter 15. These additional invariants make its hydrodynamics very peculiar (see Section 4.3). This explains the widespread use of the hexagonal model.

4.2 Hydrodynamic and microscopic scales, and the mean free path

The simple calculation we have just given illustrates the more general derivation holding for other lattice gases. Two important ingredients are apparent in this derivation. One is that we consider small deviations from thermodynamic equilibrium. The effect of collisions is to bring back the lattice towards equilibrium. This is why the eigenvalue λ must always be negative. Another ingredient is the *slow variation* of quantities such as the velocity **u** in the shear flow (4.5). This is but another expression of the separation of macroscopic and microscopic length scales discussed in Section 2.1. The velocity **u** changes significantly only over distances of the order of, or larger than the smallest hydrodynamic scale L_{hydro}. In the present case one may estimate $L_{\text{hydro}} = 1/\omega$, a large number compared to the lattice spacing, or microscopic scale, which is $\ell_{\text{lattice}} = 1$ in the HPP case. Hydrodynamics is obtained only when L_{hydro} is much larger than ℓ_{lattice}. There is however a more stringent condition.

A physical interpretation of the eigenvalue λ is the rate at which the fluid returns to equilibrium. It is easy to understand that this rate vanishes when $f \to 0$. In that limit the mean free path of a particle is long. Since for a particle on the HPP lattice in direction 1 the probability of undergoing a collision is equal to the probability of finding channels 2, 3, and 4 occupied as in Figure 3.1a, that is $p = N_3\overline{N}_2\overline{N}_4$. The mean free path is then $\ell_{\text{mfp}} = 1/p$. In the dilute gas limit ($f \to 0$) we find $p \simeq f$ and $\nu \simeq \ell_{\text{mfp}}/8$. We find that the viscosity varies as ℓ_{mfp}. This was already predicted by Maxwell who argued that

$$\nu \sim v_{th}\ell_{\mathrm{mfp}} \tag{4.33}$$

where v_{th} is the thermal velocity of molecules. Notice that as λ goes to zero with f, the mean free path, the viscosity, and the magnitude of the correction ϵ_i diverge. To keep the approximations in (4.8) and (4.16) valid, and in particular $\epsilon_i \ll N_i$, we need to keep $\omega\ell_{\mathrm{mfp}}$ small, thus we need to keep the hydrodynamic scale much larger than the mean free path as in relation (2.1).

This discussion helps us understand why one often uses models with rest particles. Let us examine the mean free path of particles. In a 7-velocity model with head-on collisions only, each moving particle has a probability $f(1-f)^5$ of colliding with another particle. (We have assumed for simplicity that all the directions have the same population f.) In addition, a moving particle has now a probability $3f(1-f)^5$ of undergoing a collision such as the one in Figure 3.2. The collision probability is much higher at relatively small f. As a result the mean free path is shorter, and convergence to hydrodynamics faster when lattice size is increased.

4.3   Navier-Stokes equation for the hexagonal model

The Boltzmann equation for the hexagonal lattice gas reads like equation (4.6) but with the new collision operator

$$\Delta_i = \left[-\tilde{N}_i\tilde{N}_{i+3} + \frac{1}{2}\left(\tilde{N}_{i+1}\tilde{N}_{i-2} + \tilde{N}_{i-1}\tilde{N}_{i+2}\right)\right. \tag{4.34}$$

$$\left. - \tilde{N}_i\tilde{N}_{i+2}\tilde{N}_{i-2} + \tilde{N}_{i-1}\tilde{N}_{i+1}\tilde{N}_{i+3}\right]\prod_j(1 - N_j) \tag{4.35}$$

where

$$\tilde{N}_i = \frac{N_i}{1 - N_i}. \tag{4.36}$$

The linearized collision operator Λ may be computed as in Section 4.1. The operator may be simplified a great deal by noting its symmetry properties. We apply invariance by $n\pi/3$ rotations and symmetry with respect to reflection about the i-axis to find that the operator is again a symmetric circulant matrix of the general form given by equation (4.21). It is convenient to write the matrix in the form

$$\Lambda = \begin{pmatrix} a_0 & a_{60} & a_{120} & a_{180} & a_{120} & a_{60} \\ a_{60} & a_0 & a_{60} & a_{120} & a_{180} & a_{120} \\ a_{120} & a_{60} & a_0 & a_{60} & a_{120} & a_{180} \\ a_{180} & a_{120} & a_{60} & a_0 & a_{60} & a_{120} \\ a_{120} & a_{180} & a_{120} & a_{60} & a_0 & a_{60} \\ a_{60} & a_{120} & a_{180} & a_{120} & a_{60} & a_0 \end{pmatrix} \qquad (4.37)$$

where the coefficients are given for the collision operator (4.35) by

$$a_0 = -f\overline{f}^2, \qquad (4.38)$$

$$a_{60} = \frac{1}{2}f\overline{f}^2(1+f), \qquad (4.39)$$

$$a_{120} = \frac{1}{2}f\overline{f}^2(1-3f), \qquad (4.40)$$

$$a_{180} = -f\overline{f}^2(1-2f). \qquad (4.41)$$

A circulant matrix of the form (4.21) has six eigenvectors $\mathbf{X}^{(0)}, \cdots, \mathbf{X}^{(5)}$ of the form $X_j^{(k)} = \xi_6^{kj}$ where $\xi_6 = \exp(i\pi/3)$ is the 6th root of unity. This may be verified by noting that

$$\lambda_k X_i^{(k)} = L_{|i-j|}X_j^{(k)} \qquad (4.42)$$

with the eigenvalues

$$\lambda_k = \sum_{\ell=0}^{5} L_\ell \xi_6^{\ell k}. \qquad (4.43)$$

These eigenvectors are physically interpreted in the following way. As in the HPP case we find three eigenvectors with zero eigenvalue. Indeed $1 = \mathbf{X}^{(0)}$ is connected to mass conservation, and the other eigenvectors are related to the real and imaginary parts of the $\mathbf{X}^{(n)}$:

$$(c_{ix})_i = \text{Re}\left\{\mathbf{X}^{(1)}\right\} = \text{Re}\left\{\mathbf{X}^{(5)}\right\}, \qquad (4.44)$$

$$(c_{iy})_i = \text{Im}\left\{\mathbf{X}^{(1)}\right\} = -\text{Im}\left\{\mathbf{X}^{(5)}\right\}, \qquad (4.45)$$

$$(Q_{ixx})_i = -(Q_{iyy})_i = \frac{1}{2}\text{Re}\left\{\mathbf{X}^{(2)}\right\} = \frac{1}{4}\begin{pmatrix} -1 \\ -1 \\ 2 \\ -1 \\ -1 \\ 2 \end{pmatrix}, \qquad (4.46)$$

and

$$(Q_{ixy})_i = \frac{1}{2}\text{Im}\left\{\mathbf{X}^{(2)}\right\} = \frac{\sqrt{3}}{4}\begin{pmatrix} 1 \\ -1 \\ 0 \\ 1 \\ -1 \\ 0 \end{pmatrix}. \tag{4.47}$$

Both eigenvectors have the same eigenvalue

$$\lambda = -3f\bar{f}^3. \tag{4.48}$$

The last eigenvector is

$$\mathbf{X}^{(3)} = \begin{pmatrix} -1 \\ 1 \\ -1 \\ 1 \\ -1 \\ 1 \end{pmatrix}. \tag{4.49}$$

Physically, the eigenvectors c_i are the components of momentum. The $Q_{i\alpha\beta}$ are non-diagonal parts of the momentum flux, or in other words the momentum flux projected orthogonally to the mass eigenvector. This decomposition into eigenvectors holds for any hexagonal model. The value of λ, however, is specific to the collision operator (4.35).

Consider the simple shear flow

$$u_x = \omega y, \tag{4.50}$$
$$u_y = 0. \tag{4.51}$$

We perform again the expansion of the lattice-Boltzmann equation as in equation (4.8). From the Fermi-Dirac equilibrium (3.42) and equation (4.50) we find

$$N_i = f\left(1 + 2\omega y c_{ix}\right) + \epsilon_i + \mathcal{O}(\omega^2). \tag{4.52}$$

Equation (4.11) then yields, at first order in ω,

$$2f\omega c_{ix}c_{iy} = \sum_j \Lambda_{ij}\epsilon_j, \tag{4.53}$$

and thus $\epsilon_i = (2f\omega/\lambda)Q_{ixy}$. Thus

$$N_i = f\left[1 + 2\omega y c_{ix} + (2\omega/\lambda)Q_{ixy}\right] + \mathcal{O}(\omega^2). \tag{4.54}$$

The new viscous stress tensor is therefore

$$\Pi_{\alpha\beta}^{\text{visc}} = \sum_i (2f\omega/\lambda)Q_{ixy}c_{i\alpha}c_{i\beta} \tag{4.55}$$

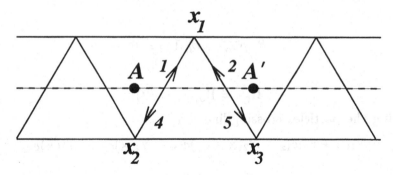

Fig. 4.2. To compute the momentum crossing the unit segment AA', one has to consider particles originating from the three sites x_1, x_2 and x_3.

and hence

$$\Pi_{xy}^{\text{visc}} = \frac{3f\omega}{2\lambda}. \tag{4.56}$$

On the other hand, in a shear flow of the form (4.50), we find from equation (4.4) that

$$\Pi_{xy}^{\text{visc}} = -\mu_1\omega. \tag{4.57}$$

Thus we find

$$\nu = -\frac{1}{4\lambda}. \tag{4.58}$$

There is however a correction to this result which we now describe.

When we estimate the momentum flux tensor as in expression (3.44) we make an error originating in the discrete nature of the lattice. To correct the error, we need first to properly account for discrete space. Thus momentum flux should be computed on a line separating the lattice sites, as in Figure 4.2. To perform this calculation correctly we need to redefine mass, momentum and fluxes *per unit area or length*. Indeed ρ in equation (3.5) is defined *per site*, but a site occupies an area of $\sqrt{3}/2$ on the lattice. (Each triangle has area $\sqrt{3}/4$, each site is bordered by six triangles, and the site shares each triangle with two other sites.) Thus per unit area we define

$$\bar{\rho} = \rho/(\sqrt{3}/2) \quad \text{and} \quad \overline{\rho\mathbf{u}} = \rho\mathbf{u}/(\sqrt{3}/2). \tag{4.59}$$

In this new framework the flux is simply the flux per unit length of the line AA' on the hexagonal lattice. In other words it is the number of particles crossing a unit length of line AA' per unit time. Since

$$\partial_t(\rho u_\alpha) + \partial_\beta\Pi_{\alpha\beta} = 0 \tag{4.60}$$

and

$$\partial_t(\overline{\rho u_\alpha}) + \partial_\beta \overline{\Pi}_{\alpha\beta} = 0, \tag{4.61}$$

we get

$$\overline{\Pi}_{\alpha\beta} = \Pi_{\alpha\beta}/(\sqrt{3}/2). \tag{4.62}$$

Counting the particles crossing line AA' ,

$$\overline{\Pi}_{xy} = N_1(\mathbf{x}+\mathbf{c}_1)c_{1x} + N_2(\mathbf{x}+\mathbf{c}_2)c_{2x} - N_4(\mathbf{x})c_{4x} - N_5(\mathbf{x})c_{5x} \tag{4.63}$$

and thus

$$\Pi_{xy} = (\sqrt{3}/2)\left[N_1(\mathbf{x}+\mathbf{c}_1)c_{1x} + N_2(\mathbf{x}+\mathbf{c}_2)c_{2x} - N_4(\mathbf{x})c_{4x} - N_5(\mathbf{x})c_{5x}\right]. \tag{4.64}$$

Substituting (4.54) we get

$$\Pi_{xy} = \rho\omega\left(\frac{1}{4\lambda} + \frac{1}{8}\right) \tag{4.65}$$

and thus comparing to (4.57) we find

$$\nu = -\frac{1}{4\lambda} - \frac{1}{8}. \tag{4.66}$$

Substituting (4.48) for λ we obtain

$$\nu = \frac{1}{12f\bar{f}^3} - \frac{1}{8}. \tag{4.67}$$

One way to measure the viscosity on the lattice is the Poiseuille viscometer method of McNamara, Kadanoff, and Zanetti. In this experiment flow is pushed through a channel at a fixed rate, adding a small amount of momentum at each site (as described in Section 7.3 below). A classical parabolic flow solution of the Navier-Stokes equations is found. Viscosity is then measured by comparing flow rate to applied forcing.

Another type of measurement may be performed using decaying shear waves. Because the shear waves eventually decay to below noise levels, such transient experiments cannot be sustained indefinitely and are in our opinion less accurate than Poiseuille viscometers. However they have the advantage that forcing is not required. A third method is to use the Green-Kubo integrals. It is then possible to estimate the viscosity from a measurement of the two-time correlations in equilibrium. This eliminates the need for forcing and for transient experiments.

Figure 4.3 shows the result of the first two methods. Measured viscosities agree within a few percent with the predicted ones. Accurate comparisions cannot be made, however, because viscosity diverges logarithmically with length in 2D. This fact is studied at length in Ref. [4.3]. Thus our derivation of viscous hydrodynamics from the Boltzmann equation is in a sense heuristic.

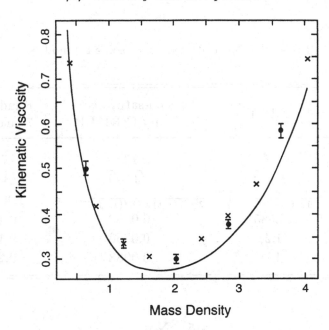

Fig. 4.3. The results of the viscosity measurements of [3.3] and [4.2]. This figure is reprinted from [4.2].

4.4 Viscosity with rest particles

Table 4.1 shows that models with rest particles have a lower viscosity. In fact, almost one order of magnitude is gained from the FHP-I model of Chapter 2 to the FHP-III or collision-saturated 7-velocity model described in Section 3.6. This is in part a consequence of the shorter mean free path.

The shear viscosity for the random 7-velocity model is not given explicitly. This is because the full analysis yields a large number of terms making the simple derivations above impractical. We thus postpone the general discussion to Chapter 15. It worthwhile however to anticipate the results of that chapter and to notice that the viscosity is still given by (4.66) where now λ is an eigenvalue of the linearized collision operator associated with an extension of the b_m-vector $Q_{i\alpha\beta}$.

In the 7-velocity case there is now a non-zero bulk viscosity (in the sense of Appendix C). This viscosity coefficient is associated specifically with uniform expansions and compressions. The presence of this viscosity adds some thermodynamic realism to the model. It also contributes to the damping of sound waves, and it may thus be responsible for a better approximation of the incompressible limit $\nabla \cdot \mathbf{u} = 0$.

Table 4.1. Viscosity values for some 2D models

	FHP-I	Collision-saturated 7-velocity (FHP-III)	Random 7-velocity
ρ	$6f$	$7f$	$7f$
c_s	$1/\sqrt{2}$	$\sqrt{3}/\sqrt{7}$	$\sqrt{3}/\sqrt{7}$
ν	$\frac{1}{12}\frac{1}{f(1-f)^3} - \frac{1}{8}$	$\frac{1}{28}\frac{1}{f(1-f)}\frac{1}{1-8f(1-f)/7} - \frac{1}{8}$	n.a.
$\nu_{f=0.3}$	0.685	0.0988	0.236
$\nu_{f=0.5}$	1.21	0.0750	0.191
$\nu_{f=0.7}$	4.28	0.0988	0.236

4.5 Notes

The analysis of the viscosity in this chapter is reminiscent of:

4.1 Hénon, M. (1987). Viscosity of a lattice gas. *Complex Systems* **1**, 763–789.

The first numerical measurements of viscosity were reported in Ref. [3.3] and in:

4.2 Kadanoff, L. P., McNamara, G. R., and Zanetti, G. (1987). A Poiseuille viscometer for lattice-gas automata. *Complex Systems* **1**, 791-803.

The divergence of viscosity in 2D is studied in:

4.3 Kadanoff, L. P., McNamara, G. R., and Zanetti, G. (1989). From automata to fluid flow: comparisons of simulation and theory. *Phys. Rev. A* **40**, 4527–4541.

The Boltzmann equation in this chapter is a discretization of an equation, valid for gases in the molecular-chaos approximation, that describes the evolution of populations $f(\mathbf{x}, \mathbf{v}, t)$ giving the number of particles per unit volume at position $\mathbf{x}$ and time t with velocity $\mathbf{v}$. One allows $\mathbf{x}$, $\mathbf{v}$ and time t to vary continuously. The equation reads

$$\partial_t f + \mathbf{v} \cdot \nabla_{\mathbf{x}} f + \frac{\mathbf{f}}{m} \cdot \nabla_{\mathbf{v}} f = \Omega(f, f) \qquad (4.68)$$

where $\nabla_{\mathbf{x}}$ is the usual differentiation (gradient) with respect to $\mathbf{x}$ and $\nabla_{\mathbf{v}}$ is differentiation with respect to $\mathbf{v}$, $\mathbf{f}$ is the force on the particle, m

the mass of the particle, and $\Omega(f, f)$ is a quadratic operator representing collisions. An introduction to the classical Boltzmann equation may be found in:

4.4 Cercignani, C., Illner, R., and Pulvirenti, M. (1994). *The Mathematical Theory of Dilute Gases* (Springer-Verlag, New York).

The Boltzmann equation of the HPP model is closely related to a class of models called *discrete velocity gases*. The idea of discretizing velocities was exploited in:

4.5 Broadwell, J. E. (1964). Shock structure in a simple discrete velocity gas. *Phys. Fluids* **7**, 1243–1247.

4.6 Harris, S. (1966). Approach to equilibrium in a moderately dense discrete velocity gas. *Phys. Fluids* **9**, 1328–1332.

4.7 Gatignol, R. (1970). Théorie cinétique d'un gaz à répartition discrète des vitesses. *Z. Flugwissenschaften* **18**, 93–97.

4.8 Hardy, J. and Pomeau, Y. (1972). Thermodynamics and hydrodynamics for a modeled fluid. *J. Math. Phys.* **13**, 1042–1051.

In the last three of these references, the idea is expressed that the discrete velocity idea could be traced back to Maxwell. Maxwell's famous paper on the kinetic theory of gases, where he derives an expression for ν, is:

4.9 Maxwell, J. (1890) On the Dynamical theory of gases, Scientific Papers II, pages 26-78 (Cambridge University Press).

However this paper contains no reference to discrete velocities. It seems that the earliest reference is an appendix on a 1D model in:

4.10 Carleman, T. (1957). *Problèmes mathématiques dans la théorie cinétique des gaz* (in French), (Scientific Publications of the Mittag-Leffler Institute, Uppsala).

Further early references may be found in:

4.11 Gatignol, R. (1975). *Théorie cinétique des gaz à répartition discrète des vitesses, Lecture Notes in Physics* **36** (Springer, Berlin).

The Boltzmann equation for the discrete velocity gas may be described as follows. Populations $N_i(\mathbf{x}, t)$ are functions of continuous time and space. There is no exclusion principle, so the $\overline{N}_i$ terms are not present. The velocity set $\{\mathbf{c}_i\}$ is still discrete. The Boltzmann equation is then

$$\partial_t N_i + \mathbf{c}_i \cdot \nabla N_i = -N_i N_{i+2} + N_{i+1} N_{i-1}. \tag{4.69}$$

The Boltzmann discrete-velocity gas and lattice-Boltzmann approaches are in a way increasingly drastic discretizations of equation (4.68). One may approach the continuous Boltzmann equation using a sufficiently large number of velocity vectors. Velocity vectors may be discretized on a lattice just as positions are. Such models with large numbers of velocities approximate the classical (continuous velocity) Boltzmann equation economically. In 3D this leads to models with 12^3 or 14^3 velocities as in:

4.12 Goldstein, D. (1991). Near-continuum applications of a discrete velocity gas model. In Beylich, A. E., editor, *Proceedings of the 17th international symposium on rarefied gas dynamics (Aachen 1990)*, pages 846–853, New York. VCH.

4.13 Inamuro, T. and Sturtervant, B. (1991). Heat transfer in a discrete-velocity gas. In Beylich, A. E., editor, *Proceedings of the 17th international symposium on rarefied gas dynamics (Aachen 1990)*, pages 854–861, New York. VCH.

A great deal of discrete-velocity gas theory carries over to lattice-gas automata. A hexagonal set of velocities $\{c_i\}$ was introduced in [4.11]. The model of Gatignol had pair collisions only, while triple collisions, which are necessary to avoid the spurious invariants discussed later (see equation (15.7)) were introduced in [1.1]. Discrete-velocity gases have undergone interesting developments on their own. For instance, evaporation and condensation have been studied recently in:

4.14 d'Almeida, A. and Gatignol, R. (1996). Evaporation and condensation problems in discrete kinetic theory. *Euro. J. Mech. B/Fluids* , in press.

5

Some simple three-dimensional models

This chapter introduces several simple lattice-gas models for 3D flow. Only simple collision rules are given, such as pair and triplet schemes and the random-collision scheme. This (short) discussion plays a mostly didactic role. It introduces lattice-gas-automata models of intermediate complexity, and prepares for the study of the 3D lattice-Boltzmann method in Chapter 6.

Another peculiarity of this chapter is its heavier use of mathematical style. This·is but an expression of the fact that the description of higher-dimensional lattices and their symmetries is a mostly geometrical subject. The mathematics, however, are rather elementary. The reader who is not familiar with the type of tensor analysis used, for instance, in the theory of the elasticity of crystal structures, should consider reading Appendix A first.

5.1   Bravais lattices

A *lattice* is any set of points in space. The lattices we use in this book are *Bravais lattices*, defined as follows:

Definition 5.1 *A Bravais lattice in D-dimensional space is defined by a set of D independent generating vectors* $\mathbf{u}_k$. *Any point of the lattice* $\mathbf{x}$ *may be written in the form of a linear combination of generating vectors with integer weights. In equations*

$$\mathbf{x} = n_1\mathbf{u}_1 + \cdots + n_D\mathbf{u}_D \qquad (5.1)$$

where all the n_i *are variable integers.*

This definition is equivalent to the following: A Bravais lattice is a set of points that *all have identical surroundings*. The Bravais lattices form a subset of, but are not equivalent to, the *periodic lattices*. Thus to

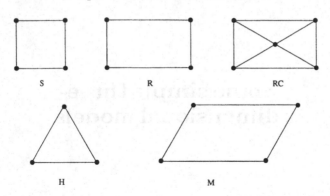

Fig. 5.1.   The five two-dimensional Bravais lattices.

describe a particular Bravais lattice it is practical to describe a fundamental pattern or cell, which, when repeated, yields the whole lattice. Many such fundamental patterns may be chosen for each type of lattice but one usually chooses patterns that display the symmetry of the lattice in an obvious way. In 2D there are altogether five Bravais lattices. One may see in Figure 5.1 fundamental cells for the square (S) , hexagonal (H), rectangular (R) and monohedral (M) lattices. In addition to these lattices, one may construct a centered rectangular lattice (RC). Out of these five lattices, only two (S and H) are useful for the construction of lattice gases. All the others lack in some respect symmetry, to the extent that one would be hard-pressed to even define a lattice-gas dynamics on such lattices.

Before we venture into three and higher dimensions, it is useful to discuss symmetry. For a start, we may remark that the S and H lattices are those for which the fundamental pattern is a regular polygon. This polygon is a highly symmetrical object. Its symmetry may be expressed by examining its symmetry group G, the group of transformations of space that leave it invariant. We specifically consider *isometries*, or *congruent transformations*. For instance, the square cell is invariant by a finite group of eight isometries, which are the identity, $\pm\pi/2$ rotations, two reflections about the Cartesian axes, two more reflections about the diagonals, and coordinate reversal.

The symmetry group that leaves a cell invariant also leaves the entire lattice invariant: it is called the *point symmetry group* of the lattice. Because of the importance of this group, it is useful, when one describes a Bravais lattice, to choose as the fundamental cell a set of points that is invariant under G. Indeed that is what we did in our description of the 2D Bravais lattices.

One interesting fundamental cell is formed by a set of neighbors P_i of a fixed point O of the lattice. By our previous remark we take a set that is invariant under $\mathcal{G}$. When we define a lattice gas model, we shall take as *velocity vectors*

$$\mathbf{c}_i = \vec{OP}_i. \tag{5.2}$$

Whatever the point O we can define the same set of neighbors. This is a consequence of the property of Bravais lattices that all the points have identical surroundings.

Our next step is to list symmetry properties that may or may not be satisfied by a particular Bravais lattice. These properties will play an important role in the remainder of this book. We shall list some properties that we shall not immediately study in detail for all lattices, and others which are discussed at more length in Appendix A. However we have them all here since it is useful to have all the relevant facts in one place, and a preliminary discussion of symmetry is anyway indispensable at this stage.

First we notice that for all Bravais lattices we have

Property 5.1 *For each vector* $\mathbf{c}_i$, *there is a vector* $\mathbf{c}_j = -\mathbf{c}_i$.

To prove this property, notice that the definition of a Bravais lattice implies that $\mathcal{G}$ always contains a transformation that reverses all coordinates. Thus any fundamental cell must be invariant by coordinate reversal.

Another natural property is

Property 5.2 *All vectors* $\mathbf{c}_i$ *have equal norm.*

This is true for some Bravais lattices only. To further analyze symmetry, we often need to consider tensors $\mathbf{C}^{(r)}$ formed with the vertices by

$$C^{(r)}_{\alpha_1\ldots\alpha_r} = \sum_i c_{i\alpha_1}\ldots c_{i\alpha_r}. \tag{5.3}$$

These tensors are but a special case of tensors invariant by all the lattice symmetries in $\mathcal{G}$ and symmetric in the four indices $\alpha\beta\gamma\delta$. More generally, we may consider tensors such as $\lambda_{\alpha\beta\gamma\delta}$ in Chapter 2 or the tensor $\tilde{\mu}_{\alpha\beta\gamma\delta}$ in Chapter 4 that are invariant under all the lattice symmetries in $\mathcal{G}$ but symmetric only under the exchange of $\alpha\beta$ and $\gamma\delta$. For a given lattice, it is important to determine whether tensors have any of the following two properties:

Property 5.3 *All tensors up to fourth order and invariant under all the lattice symmetries in* $\mathcal{G}$ *have cubic symmetry.*

Property 5.4 *All tensors up to fourth order and invariant under all the lattice symmetries in* $\mathcal{G}$ *have spherical symmetry, i.e. they are isotropic.*

Table 5.1. Symmetry properties of lattices

Lattice	Square	Hexagonal	Cubic	Centered Cubic	Face-Centered Cubic	Face-Centered Hypercubic
Symbol	S	H	C	CC	FCC	FCHC
Polytope	square	hexagon	octahedron	cube	(irregular)	$\{3,4,3\}^*$ or P_{24}
Symmetry group	D_4	D_6	$O_h^\dagger$	$O_h^\dagger$	$O_h^\dagger$	group of $\{3,4,3\}^*$
Isotropic fourth order tensors	no	yes	no	no	no	yes
Number of vectors c_i	4	6	6	8	12	24
Number of elements in group‡	8	12	48	48	48	1152

* See Appendix B for a definition of this *Schläfli symbol*.
† The *octahedral group* or group of cubic symmetry.
‡ See Appendix B for a calculation of that number.

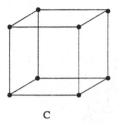 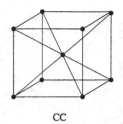 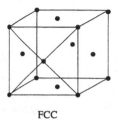

C CC FCC

Fig. 5.2. The three cubic lattices. C is simple cubic, CC is centered cubic and FCC is face centered cubic.

These two symmetry properties require some discussion. Cubic symmetry is defined in Appendix A, where we also show that it is shared by all the lattices in Table 5.1. There we also derive the classical result that Property 5.4 is true for the H lattice.

Among 3D lattices, we find 14 Bravais lattices, of which only 3 have cubic symmetry. These three lattices are the simple cubic (C), centered cubic (CC) and face centered cubic (FCC) lattices. They are shown in Figure 5.2. As implied by their names, they all have Property 5.3. However, as Table 5.1 shows, *none* has Property 5.4, that is isotropic fourth-order tensors. However, as we have seen in Chapters 2 and 4, this property is important in deriving an isotropic Euler equation as well as isotropic viscous tensors. The FCHC lattice offers an interesting procedure to circumvent this difficulty.

5.2   The FCHC–projection models

As Table 5.1 shows, there are no lattices in 3D that have the required symmetry properties, specifically Property 5.4. A way around this difficulty is to make a detour through the fourth dimension. The FCHC–projection models have 24 bits n_i and 24 velocity directions c_i of the form

$$c_i = \text{perm}(\pm 1, \pm 1, 0, 0) \tag{5.4}$$

where the notation $\text{perm}(a, b, c, d)$ indicates all the vectors obtained by permutations of the symbols a, b, c, d. There are six permutations and four ways to choose the signs and thus 24 vectors/vertices. The 24 vectors c_i form a 4D regular object called the 24-cell *polytope* P_{24} (see Figure 5.3). Besides being of obvious mathematical beauty, this polytope has the distinct advantage of having a large enough symmetry group. As a result the FCHC lattice has Property 5.4.

Despite the 4D nature of the polytope, we do not really work in 4D, but instead we adapt this object to construct a lattice gas model on plain 2D and 3D cubic lattices. There is a simple correspondence between the

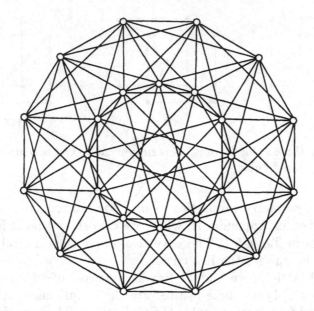

Fig. 5.3.   A projection of the 24-vertex polytope made of the nearest neighbors in the FCHC lattice [5.3]. The 24 vertices are shown as small circles. Each face has three edges. The eight edges attached to each vertex join it to a cube.

24 velocities of FCHC and a fundamental pattern of the cubic lattice invoving first and *second* neighbors. For instance, if we consider only the first three components of c_i we obtain twelve vectors like

$$\mathbf{u} = \mathrm{perm}(\pm 1, \pm 1, 0) \tag{5.5}$$

which lie entirely on diagonals of a 3D cubic lattice and connect second neighbors. There are six other vectors, like

$$\mathbf{v} = \mathrm{perm}(\pm 1, 0, 0) \tag{5.6}$$

which connect nearest neighbors of the 3D cubic lattice. These vectors each correspond to a pair of 4D vectors. For instance, $\mathbf{v} = (1, 0, 0)$ is the projection of

$$\mathbf{v}' = (1, 0, 0, 1) \quad \text{and} \quad \mathbf{v}'' = (1, 0, 0, -1). \tag{5.7}$$

The projected FCHC model has thus *two* velocity vectors and *two* types of particles connecting nearest neighbors. The situation is summarized in Figure 5.4.

   If we project the FCHC lattice into 2D we find four particles with velocity $(0, 0)$, 16 joining nearest neighbors (with velocity of the type $\mathbf{v} = \mathrm{perm}(\pm 1, 0)$), and four on the diagonals, with velocity of the type $\mathbf{u} = \mathrm{perm}(\pm 1, \pm 1)$. The situation is summarized by Figure 5.5.

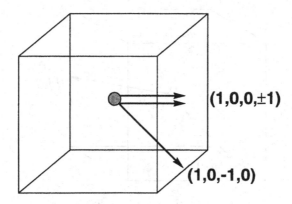

Fig. 5.4. A perspective view of the FCHC primitive cell, projected into 3D space. Instead of explicitly showing all 24 velocities, only two of the twelve velocities which extend into the fourth dimension are shown, along with just one of the velocities with no component in the fourth dimension.

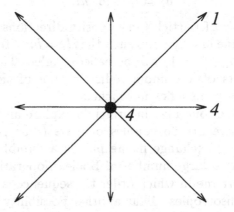

Fig. 5.5. A projection of the FCHC lattice on a 2D square lattice. The numbers indicate the number of vectors in 4D that project on the 2D vector.

5.3 Collision operators

Our first scheme consists of a list of elementary collisions involving two or three particles, which we call the *pair and triplet scheme*. We start by labelling the 24 directions c_i in such a way that

$$c_{i+12} = -c_i, \tag{5.8}$$

and then decompose the 24–bit state vector n_i into two twelve-bit components $l_i = n_i$ for $0 \leq i < 12$ and $h_i = n_{i-12}$ for $12 \leq i < 24$. We then

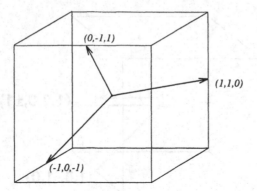

Fig. 5.6. An example of a velocity triplet that sums to zero. This triplet is entirely contained in 3D, but may be considered either as an element of the FCC set of velocity vectors, or as an element of the FCHC set.

define the dumbbell variable

$$d_i = l_i \text{ and } h_i. \tag{5.9}$$

A dumbbell is a pair of particles in opposite directions corresponding to $d_i = 1$. We also define empty pairs such that $l_i = h_i = 0$. Pairs containing a unique particle ($h_i + l_i = 1$) will be called singles. The first step of the pair scheme consists of a random redistribution of the dumbbells and empty pairs, leaving the singles untouched.

In computer coding, one may in principle explore all the pairs c_i, c_{i+12} in a sequence. There are 66 couples of pairs $(c_i, c_{i+12}, c_j, c_{j+12})$ with $i \neq j$, which we may exchange if one pair is a dumbbell and the other empty. This involves a large number of Boolean operations. Moreover, it is necessary to determine in which order the sequence is investigated, and this may involve anisotropies. Thus another possibility is to "empty" all the dumbbells, then redistribute the particles in pairs among the resulting set of empty pairs. This redistribution is most simply performed at random.

In 3D and 4D there are other pair collisions that involve non-opposite pairs. We shall give examples in the purely 3D case of the twelve FCC vectors

$$\mathbf{c}_i = \text{perm}(\pm 1, \pm 1, 0). \tag{5.10}$$

Clearly the FCC vertices are a 3D subset of the FCHC vertices, and the collisions we shall find may easily be extended to 4D. Consider pairs of particles that have total momentum $\mathbf{g} = (2, 0, 0)$, for instance

$$\mathbf{c}_1 = (1, 1, 0) \tag{5.11}$$
$$\mathbf{c}_2 = (1, -1, 0). \tag{5.12}$$

(The labels 1, 2 in this example bear no connection with the labels defined previously.) Clearly $c_1 + c_2 = g$. There are two pairs adding to g in FCC, while in FCHC there are three pairs adding to $g = (2, 0, 0, 0)$. This makes one type of collision in FCC and three in FCHC (a collision is defined as a pair of pairs). If we consider pairs adding to all the momenta $g = \text{perm}(\pm 2, 0, 0)$ we get six values of g and as many possible collisions in FCC. In FCHC we consider $g = \text{perm}(\pm 2, 0, 0, 0)$, there are eight values of g which makes 24 collisions. Another type of collisions is obtained from the set of pairs that add to $g = \text{perm}(\pm 1, \pm 1, 0)$. An example is

$$c_1 = (1, 0, 1) \tag{5.13}$$
$$c_2 = (0, 1, -1). \tag{5.14}$$

We find for each g two pairs in FCC and four in FCHC.

This procedure may be generalized through the idea of local mass-momentum packets defined in Section 3.6. There are only three types of mass-momentum packets with two particles which are degenerate, that is those that involve more than one pair. Thus we have exhausted all the pair collisions in FCC and FCHC.

As a final refinement, one may consider collisions with three particles. Here we will not list all the mass-momentum packets with three particles, but consider only those of the form $(m, g) = (3, 0)$. These packets contain three velocities c_1, c_2, and c_3 that sum to zero, such as for instance:

$$c_1 = (1, 1, 0) \tag{5.15}$$
$$c_2 = (-1, 0, 1) \tag{5.16}$$
$$c_3 = (0, -1, -1). \tag{5.17}$$

There are eight such triplets of velocities in FCC and 32 in FCHC. If we look at a cubic cell from the direction of one of the diagonals of the cube, c_1, c_2, and c_3 form the symmetric pattern shown in Figure 5.6. The four diagonals of the cube support each two such triplets yielding eight such triplets of velocities in FCC. A collision may then exchange any full triplet with any empty one. Momentum is unchanged since it is zero for the full triplets as well as the empty ones. To decide which triplets are swapped it is possible to use a random number generator.

So far, in the pair and triplet collisions we have used randomness only as a convenience, to chose between some competing collisions. A much more drastic use of randomness is made in the so called *random scheme*. There output states are chosen at random among all the states in the same mass-momentum packets as the input state. A typical computer implementation of the random scheme can be given as follows.

1. The mass and momentum of a site are computed.

2. This calculation is then translated, through an appropriate table into a pointer $T^{(1)}$.

3. This pointer $T^{(1)}$ points to a table $T^{(2)}_{(m,g)}$ containing all the configurations in the packet. An element of this table is then picked with the help of a random number generator.

The resulting scheme is quite general, and could just as well be used in FHP or other lattices.

Both the pair and triplet scheme and the random scheme suffer from a relatively large viscosity, and a rather long mean free path. This may cause problems in some applications, but the models are attractively simple and form the basis of a number of models studied in practice. As a quick fix, it is possible to add rest particles. The interested reader may try the exercises below.

5.4   Exercises

5.1 Find the group of isometries of the hexagonal cell.

5.2 Find the polytopes constructed with the nearest neighbors of the vertices in each of the three cubic lattices.

5.3 Find the number of triplets in FCHC.

5.4 Find a collision on the FCC and FCHC lattice involving a rest particle. (Hint: use the triplets of section 5.3.)

5.5 How does the introduction of rest particles in FCC affect the mean free path?

5.5   Notes

The early years of lattice gas automata research saw a frantic scramble for the development of a 3D equivalent of the hexagonal lattice: a lattice on which Property 5.4 would hold. Lattices up to high dimensionality were studied by computations involving group characters in:

5.1 Wolfram, S. (1986). Cellular automaton fluids 1: Basic theory. *J. Stat. Phys.* **45**, 471–526.

The now widely used idea of a projected FCHC lattice was first published in:

5.2 d'Humières, D., Lallemand, P. and Frisch, U. (1986). Lattice gas models for 3D hydrodynamics, *Europhys. Lett.* **2**, 291–297.

The detailed geometrical study of polytopes such as P_{24} is given in:

5.3 Coxeter, H. S. M. (1977). *Regular Polytopes* (Dover).

In practice, models much more complex than those described here have been used. These models are still defined on the FCHC lattice, with in some cases one or more rest particles. Two issues are relevant when designing them. One is to select a collision operator, and historically, the emphasis has been on reducing the viscosity of the resulting model. The other is to find a compact description, using tables as small as possible. A series of FCHC models of ever-decreasing viscosity were defined in:

5.4 Hénon, M. (1989). Optimization of collision rules in the FCHC lattice gas, and addition of rest particles. In Monaco, R., editor, *Discrete Kinetic Theory, Lattice-Gas Dynamics, and Foundations of Hydrodynamics*, pages 146–159 (World Scientific, Singapore).

5.5 Dubrulle, B., Frisch, U., Hénon, M., and Rivet, J.-P. (1990). Low viscosity lattice gases. *J. Stat. Phys.* **59**, 1187–1226.

5.6 Hénon, M. (1992). Implementation of the FCHC lattice gas model on the Connection Machine. *J. Stat. Phys.* **68**, 353–377.

However the issue of low viscosity is irrelevant for low-Reynolds-number hydrodynamics. There it seems that one should try to optimize the mean free path, or some combination of eigenvalues of the collision operator. Since to some approximation viscosity varies as the inverse of the mean free path, as in Maxwell's formula (4.33), its minimization amounts approximately to the minimization of the mean free path.

Isometries and mass-momentum packets are useful when building algorithms. Isometries are used to map equivalent configurations on each other. They act on the local lattice-gas configurations as a subset of the group of permutations of the 24 bits $(n_i)_i$. To generate these permutations, it is important to remark that each of the 1152 isometries in the symmetry group of P_{24} may be decomposed uniquely into a product of elementary isometries, of the form:

$$\begin{pmatrix} \mathbf{I} \\ \mathbf{S}_1 \\ \mathbf{S}_2 \end{pmatrix} \begin{pmatrix} \mathbf{I} \\ \mathbf{P}_1 \end{pmatrix} \begin{pmatrix} \mathbf{I} \\ \mathbf{P}_2 \end{pmatrix} \begin{pmatrix} \mathbf{I} \\ \mathbf{P}_3 \end{pmatrix} \begin{pmatrix} \mathbf{I} \\ \mathbf{P}_4 \end{pmatrix} \begin{pmatrix} \mathbf{I} \\ \mathbf{P}_{12} \end{pmatrix} \begin{pmatrix} \mathbf{I} \\ \mathbf{P}_{13} \\ \mathbf{P}_{23} \end{pmatrix} \begin{pmatrix} \mathbf{I} \\ \mathbf{P}_{14} \\ \mathbf{P}_{24} \\ \mathbf{P}_{34} \end{pmatrix}. \quad (5.18)$$

The meaning of the notations and the proof of the decomposition are given in Appendix B.

In the FCHC case, mass-momentum packets are much larger and more numerous than in 2D. However they may be tabulated numerically without much trouble. There are 25 possible values of the number of particles m and 7009 possible values of $\mathbf{g}$. However, some values of the momentum are equivalent to each other in the sense that they can be transformed into each other by isometries. There are only 37 values of the momentum that are not equivalent. Another simplification is obtained if we consider particle-hole symmetry (the transformation $s_i \rightarrow 1 - s_i$). This reduces the number of mass values to 13. Finally a number of packets $(m, \mathbf{g})$ are empty or contain a single state. They cannot contain a collision. With these simplifications, Hénon [5.4] arrived at only 154 relevant packets, the size $|(m, \mathbf{g})|$ of each packet being at most 8952.

Since the collision operator sends mass-momentum packets onto themselves, it factorizes over the packets, and the operator selection problem reduces to the definition of the operator over each packet independently. Using these and similar ideas, Rem and Somers arrived at an algorithm that satisfies semi-detailed balance (defined in Chapter 14) and is optimized. It uses several tables fitting together in less than 120 kbytes of memory. It is described in:

5.7 Somers, J. A. and Rem, P. C. (1992). Obtaining numerical results from the 3D FCHC-lattice gas. In *Springer Proceedings on Physics, Workshop on Numerical Methods for the Simulation of Multi-Phase and Complex Flow*, pages 59–78 (Springer-Verlag, Berlin).

A less optimized, but simpler collision operator may be found in:

5.8 Adler, C., Boghosian, B. M., Flekkøy, E. G., Margolus, N. H., and Rothman, D. (1995). Simulating three-dimensional hydrodynamics on a cellular-automata machine. *J. Stat. Phys.* 81, 105–128.

Work on 3D models is clearly unfinished. Since viscosity is not the appropriate parameter to optimize, tables should be reconstructed with more relevant optimization criteria, such as minimizing the mean free path. Other future developments may follow the trend for models using finite precision (integer) arithmetic such as those in:

5.9 Boghosian, B. M., Yepez, J., Alexander, F. J., and Margolus, N. H. (1996). Integer lattice gases. preprint `comp-gas/9602001` at the electronic archive `http://xyz.lanl.gov`.

6

The lattice-Boltzmann method

The lattice-Boltzmann method is a technique for the numerical solution of the incompressible Navier-Stokes equation. Although it is closely related to the kinetic theory of (Boolean) lattice-gas cellular automata for which the *Boltzmann approximation* has been defined in Chapter 4, we view it primarily as a tool of computational fluid dynamics. In this chapter we define a simple version of the method, called the single-relaxation-time model. We give a didactic explanation of how the lattice-Boltzmann method produces an approximation to the solutions of the Navier-Stokes equation. The full derivation of the Navier-Stokes equation is given later in Chapter 15.

The presentation of the method in this chapter is limited to the central issue of bulk hydrodynamic behavior. Other important issues, such as boundary conditions and stability, are discussed in Chapter 7.

6.1 Basic definitions

To start with, as in Boolean lattice gases, we consider the hexagonal lattice in 2D or the FCHC lattice projected to 2D or 3D. We have a set of population densities, written $N_i(\mathbf{x}, t)$, where $\mathbf{x}$ is a lattice site and t is discrete time. The population variables N_i are in theory arbitrary real numbers, and are represented as floating point numbers in computer codes. As in Section 3.7 we have b_m velocity vectors ($b_m = 6$ on the hexagonal lattice and $b_m = 24$ on the FCHC lattice) noted $\mathbf{c}_i$. The velocity modulus is $c = |\mathbf{c}_i|$ for $i > 0$, with $c = 1$ for FHP and $c = \sqrt{2}$ for FCHC. We will allow for rest particles. Their total density will be noted $N_0(\mathbf{x}, t)$ and their velocity $\mathbf{c}_0 = 0$. Mass and momentum are defined as usual:

$$\rho = N_0 + \sum_{i>0} N_i \tag{6.1}$$

73

and

$$\rho\mathbf{u} = \sum_i N_i \mathbf{c}_i. \tag{6.2}$$

Notice that, contrary to the definition in equation (3.50), N_0 contains the total mass at rest and not a fraction $1/b_r$ of it.

The lattice-Boltzmann method consists of the numerical integration of an equation like (4.6). A particular case of great interest is the *pseudo-equilibrium collision operator*, defined as follows. Consider first the pseudo-equilibrium distributions

$$N_i^{(P)} = f \left[1 + \frac{c_{i\alpha} u_\alpha}{c_s^2} + \left(G_m Q_{i\alpha\beta} + G_c c^2 \delta_{\alpha\beta} \right) u_\alpha u_\beta \right] \tag{6.3}$$

and

$$N_0^{(P)} = f(b_c - b_m G_c c^2 u^2) \tag{6.4}$$

where $f = \rho/b$, $b = b_m + b_c$, $c_s^2 = b_m c^2/(bD)$, $Q_{i\alpha\beta}$ is defined as in equation (3.43), and b_c, G_m and G_c are adjustable parameters. The expressions (6.3) and (6.4) are somewhat familiar: they bear a formal resemblance with the rest particle low velocity Fermi-Dirac distributions (3.52) and (3.53). The motivation for the small differences with the true rest particle distributions will appear in Section 6.2.

It is easy to check that the pseudo-equilibrium population has the same invariants as the original populations:

$$\rho = N_0^{(P)} + \sum_{i>0} N_i^{(P)} \tag{6.5}$$

and

$$\rho\mathbf{u} = \sum_i N_i^{(P)} \mathbf{c}_i. \tag{6.6}$$

With these definitions, the lattice-Boltzmann equation with pseudo-equilibrium operator is

$$N_i(\mathbf{x} + \mathbf{c}_i, t + 1) - N_i(\mathbf{x}, t) = \sum_j \Lambda_{ij} \left[N_j(\mathbf{x}, t) - N_j^{(P)}(\mathbf{x}, t) \right]. \tag{6.7}$$

where Λ_{ij} is a matrix of constant coefficients. In essence, equation (6.7) is a simplification of the true Boltzmann operator (4.6) amounting to a linearization around the equilibrium state.

As we have seen in Chapter 4, it is the return to equilibrium produced by the Boltzmann operator that causes the viscous effects. This idea of return to equilibrium may lead us to simplify the equations one step further: we replace the linearized operator Λ_{ij} by a diagonal operator with a single eigenvalue λ. The resulting *single relaxation time lattice-Boltzmann model* (or *single relaxation model*) reads

$$N_i(\mathbf{x} + \mathbf{c}_i, t + 1) = (1 + \lambda)N_i(\mathbf{x}, t) - \lambda N_i^{(P)}(\mathbf{x}, t). \qquad (6.8)$$

The main properties of this construction are (i) mass and momentum are conserved and (ii) the populations N_i, in the absence of external forcing, *converge to the pseudo-equilibrium values* $N_i^{(P)}$. The typical time of convergence, or relaxation time of the method is $1/\lambda$. This method is a direct analogue of a similar model for the continuous space Boltzmann equation, discussed in the Notes section at the end of this chapter. To clarify matters, it is useful to formulate the lattice-Boltzmann method as a step by step numerical algorithm, in the following way:

1. As a starting point, the populations $N_i(\mathbf{x}, t)$ arriving on a site are given.

2. The mass and momentum are calculated using (6.1) and (6.2).

3. The pseudo-equilibrium populations $N_i^{(P)}$ are computed.

4. In a *collision step* analogous to the lattice-gas collision step, post-collision populations N_i' are computed :

$$N_i'(\mathbf{x}, t) = (1 + \lambda)N_i(\mathbf{x}, t) - \lambda N_i^{(P)}(\mathbf{x}, t). \qquad (6.9)$$

5. In a *streaming step* populations are shifted around the lattice:

$$N_i(\mathbf{x} + \mathbf{c}_i, t + 1) = N_i'(\mathbf{x}, t). \qquad (6.10)$$

This procedure involves communication of information between sites only in step 5. It is thus a rather simple scheme, somewhat akin to explicit finite-difference schemes. We end this section by noting that yet another, one-line formulation of the method is given in Exercise 6.6.

To show how the lattice-Boltzmann method compares with the lattice-gas, we show an example of a flow behind a cylinder. The cylinder is located in the middle of a channel. Figure 6.1a shows the result of a lattice-gas calculation. Velocity vectors are plotted. Clearly, the velocity field is quite noisy. Figure 6.1b shows the result of a lattice-Boltzmann relaxation model calculation. The velocity field is considerably less noisy.

6.2 Euler equation

To illustrate the lattice-Boltzmann method, we show how it can approach the classical *incompressible* flow equations. These equations are obtained

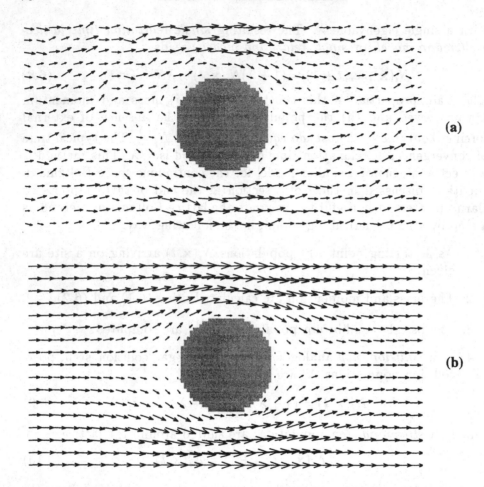

Fig. 6.1. Flow past a cylinder: (a) Lattice-gas calculation. (b) Boltzmann method.

by setting ρ to a constant ρ_0 in the mass and momentum equations (2.7) and (2.13). To avoid confusion with the lattice-gas velocity, we note the velocity field in the classical Euler equations $\mathbf{v}(\mathbf{x}, t)$. Then one obtains

$$\partial_t \mathbf{v} + \mathbf{v} \cdot \nabla \mathbf{v} = -\nabla \omega, \tag{6.11}$$

where the rescaled pressure is $\omega = p/\rho_0$. The continuity equation becomes

$$\nabla \cdot \mathbf{v} = 0. \tag{6.12}$$

To see how these equations are approached in the Boltzmann method we first remark that the relaxation to equilibrium forces the populations to approach their pseudo-equilibrium values:

$$N_i(\mathbf{x}, t) \simeq N_i^{(P)}(\mathbf{x}, t). \tag{6.13}$$

The momentum-flux tensor may be then be calculated approximately. We call $\Pi^{(P)}$ the resulting approximation:

$$\Pi^{(P)}_{\alpha\beta} = \sum_i N^{(P)}_i c_{i\alpha} c_{i\beta}. \tag{6.14}$$

Then from (6.3), (6.4) and (6.13), the momentum balance (2.50) becomes

$$\partial_t(\rho u_\alpha) + \partial_\beta(g(\rho)\rho u_\alpha u_\beta) = -\partial_\alpha[p(\rho, u^2)] \tag{6.15}$$

where

$$g(\rho) = \frac{b_m}{b} \frac{2c^4}{D(D+2)} G_m \tag{6.16}$$

and

$$p(\rho, u^2) = c_s^2 \rho \left[1 + \left(G_c c^2 - G_m \frac{2c^2}{D(D+2)} \right) u^2 \right]. \tag{6.17}$$

The coefficients G_m and G_c may be chosen to obtain the desired equations. An interesting choice is

$$G_c = \frac{\rho}{c^4} \frac{b}{b_m}, \qquad G_m = \frac{D(D+2)}{2c^4} \frac{b}{b_m} \rho. \tag{6.18}$$

Then $g(\rho) = \rho$ and p is independent of u. The reason we like $g(\rho) = \rho$ and not $g(\rho) = 1$ appears when we make the change of variable

$$\mathbf{v} = \rho \mathbf{u}. \tag{6.19}$$

Then the new momentum balance equation is

$$\partial_t v_\alpha + \partial_\beta(v_\alpha v_\beta) = -\partial_\alpha \omega, \tag{6.20}$$

where the pressure ω is related to the density by

$$\omega = c_s^2 \rho. \tag{6.21}$$

It is now clear that $c_s^2 = \partial\omega/\partial\rho$ is the soundspeed squared. Equation (6.20) may also be written

$$\partial_t \mathbf{v} + \mathbf{v} \cdot \nabla \mathbf{v} + (\text{div } \mathbf{v})\mathbf{v} = -\nabla \omega. \tag{6.22}$$

This equation is coupled with the mass conservation equation

$$\partial_t \rho + \text{div } \mathbf{v} = 0. \tag{6.23}$$

These equations differ from (6.11) and (6.12), but the difference vanishes when $\partial_t \rho = 0$.

Thus in *stationary* flow, the lattice-Boltzmann method provides the correct equations without further approximations. This is the reason why we set $g(\rho) = \rho$. The appearance of correct stationary flow is a marked advantage over the Boolean lattice-gas approach which requires low Mach numbers.

For time-dependent flow equations (6.22) and (6.23) approach the incompressible flow equations (2.13). To summarize the discussion in Appendix D let us define a typical velocity scale V (for instance the maximum velocity in the flow). Then the Mach number is $M = V/c_s$. When the Mach number is small, it may be shown that there are solutions of equations (6.22) and (6.23) that converge to incompressible solutions.

In ending this section we notice that one coefficient, b_c, remains undetermined. Stability considerations given in the following chapter result in an optimal value $b_c = 4$ for the hexagonal lattice.

The best approach to the incompressible Navier-Stokes equation is probably the one outlined above. There are however other possibilities which are instructive to consider. One such approach is outlined in Exercise 6.2.

6.3   Viscous flow

As in Chapters 3 and 4, isotropy of 4th order tensors (Property 5.4) ensures that the viscous terms have the correct form. Neglecting compressibility effects we have

$$\partial_t \mathbf{v} + \mathbf{v} \cdot \nabla \mathbf{v} = -\nabla \omega + \nu \nabla^2 \mathbf{v}, \tag{6.24}$$

where ν is the kinematic viscosity, related to the eigenvalue λ by

$$\nu = \frac{c^2}{D+2} \left(-\frac{1}{\lambda} - \frac{1}{2} \right). \tag{6.25}$$

The full compressible Navier-Stokes equation and the above expression for ν are derived in Chapter 15.

Another interesting case is the Stokes equation, which is obtained when the left-hand side of equation (6.24) can be neglected. In this case

$$0 = -\nabla \omega + \nu \nabla^2 \mathbf{v}. \tag{6.26}$$

If one sets $G_m = G_c = 0$ in (6.3) and (6.4), one gets

$$\partial_t \mathbf{v} = -\nabla \omega + \nu \nabla^2 \mathbf{v}. \tag{6.27}$$

For stationary flow, or slow time dependence, the Stokes equation (6.26) is recovered. Interestingly, with $G_m = G_c = 0$ the lattice-Boltzmann equation (6.7) simplifies and becomes truly linear:

$$N_i(\mathbf{x} + \mathbf{c}_i, t + 1) = A_{ij} N_j(\mathbf{x}, t). \tag{6.28}$$

Here A_{ij} is a matrix that may be computed from the definitions (6.4), (6.5) and (6.6).

6.4   Exercises

6.1 Show that the distributions $N_i^{(P)}$ obey the constraints on mass and momentum.

6.2 Show that with an appropriate choice of parameters G_m and G_c, the following equations may be obtained:

$$\partial_t \mathbf{u} + \mathbf{u} \cdot \nabla \mathbf{u} = -\frac{1}{\rho} \nabla p, \qquad (6.29)$$

where $p = c_s^2 \rho$ and

$$\partial_t \rho + \mathrm{div}\,(\rho \mathbf{u}) = 0. \qquad (6.30)$$

6.3 In a polytropic gas (see Appendix A) the pressure is given by $p = A\rho^\gamma$ where A and γ are given constants. How could the compressible flow of a polytropic gas be modeled?

6.4 It is interesting to compare the advantages of the approach in the text and the approach in Exercise 6.2. Discuss the following properties of the resulting macroscopic equations: (i) What quantities are conserved in the equations? (ii) What version of the equations has Galilean invariance? (Use the discussion of Galilean invariance in Chapter 2.)

6.5 One may easily add a force term in the momentum balance equation, by a simple redistribution of the populations that adds momentum to each site at a given rate (see Section 7.3). Consider a flow forced by a gravity field $\mathbf{f} = -g\hat{\mathbf{z}}$. Find the steady solution for this flow. Find $\mathrm{div}\,\mathbf{u}$ for the model in exercise 6.2. What does the result tell you about the relative merits of the two models?

6.6 Show that the lattice-Boltzmann method may be rewritten

$$N_i(\mathbf{x} + \mathbf{c}_i, t + 1) = A_{ij} N_j(\mathbf{x}, t) + B_{ijk} N_j(\mathbf{x}, t) N_k(\mathbf{x}, t). \qquad (6.31)$$

Express the coefficients A_{ij} and B_{ijk} assuming given parameters G_m, G_c.

6.5   Notes

Reviews of lattice-Boltzmann methods may be found in:

6.1 Benzi, R., Succi, S., and Vergassola, M. (1992). The lattice Boltzmann equation: Theory and applications. *Phys. Rep.* **222**, 145–197.

6.2 Qian, Y. H., Succi, S., and Orszag, S. A. (1995). Recent advances in lattice Boltzmann computing. In Stauffer, D., editor, *Ann. Rev. of Comput. Phys. III*, pages 195–242 (World Scientific, Singapore).

A recent comparison with other numerical methods and experimental work may be found in:

6.3 Qian, Y. H., Succi, S., Massaioli, F. and Orszag, S. A. (1996). A benchmark for lattice BGK model: flow over a backward-facing step. In Lawniczak, A. and Kapral, R., editors, *Pattern Formation and Lattice Gas Automata, Fields Institute Communications* 6, pages 207–215 (American Mathematical Society, Providence).

The lattice-Boltzmann method described in this chapter is the result of successive simplifications performed by several groups. McNamara and Zanetti studied the discrete lattice-Boltzmann equation obtained directly from the Boolean models, i.e. with the collision operator (4.35), in:

6.4 McNamara, G. and Zanetti, G. (1988). Use of the Boltzmann equation to simulate lattice-gas automata. *Phys. Rev. Lett.* **61**, 2332–2335.

The use of a *linearized* collision operator was suggested in:

6.5 Higuera, F. and Jimenez, J. (1989). Boltzmann approach to lattice gas simulations. *Europhys. Lett.* **9**, 663–668.

Linearization means that the lattice-Boltzmann equation is expression (6.7) and the operator Λ is given by equation (4.37) with *fixed* coefficients a_i. It is then possible to invert equation (4.42) to obtain the values of the a_i as a function of the desired values of the eigenvalues. This may be extended to 3D, with more complex forms for the linearized operator, as done in:

6.6 Higuera, F. and Succi, S. (1989). Simulating the flow around a cylinder with a lattice-Boltzmann equation. *Europhys. Lett.* **8**, 517–521.

6.7 Higuera, F., Succi, S., and Benzi, R. (1989). Lattice gas dynamics with enhanced collisions. *Europhys. Lett.* **9**, 345–349.

The idea of having a single relaxation time, and hence a diagonal Λ, originates in:

6.8 Qian, Y., d'Humières, D., and Lallemand, P. (1992). Lattice BGK models for Navier-Stokes equation. *Europhys. Lett.* **17**, 479–484.

6.9 Chen, H., Chen, S., and Matthaeus, W. H. (1992). Recovery of Navier-Stokes equations using a lattice-gas Boltzmann method. *Physical Review A* **45**, R5339–R5342.

Yet another model is obtained if one removes the exclusion principle. Taking as an example the HPP model, this yields

$$N_i(\mathbf{x} + \mathbf{c}_i, t + 1) - N_i(\mathbf{x}, t) = -N_i N_{i+2} + N_{i+1} N_{i-1}. \tag{6.32}$$

The equilibrium distributions (barring spurious invariants) for such models are the same as for discrete velocity gases: they are the Maxwellian distributions

$$N_i^{eq} = \frac{1}{Z} \exp(-h - \mathbf{q} \cdot \mathbf{c}_i) \tag{6.33}$$

with Z a normalization constant.

7

Using the Boltzmann method

In this chapter we discuss a number of practical issues that arise when performing a numerical simulation with the lattice-Boltzmann method. We first discuss the rescaling of the method to adjust to the real dimensions of the flow. This is mathematically trivial but of great importance in practice. Then we discuss two of the most interesting issues regarding the method: its stability and convergence properties. Finally we turn to forcing and boundary conditions.

7.1   Rescaling to physical variables

In applications of a numerical method we think of a problem with given boundary conditions, initial conditions and applied external forces. The boundary conditions may involve a specific geometry, with objects of given sizes, entry and exit conditions, etc., as in the flows of Section 6.1. These *data* on the problem come first and the method comes next. Thus it is impractical to continue to define *a priori* the lattice spacing as 1 or c. One should rather let the lattice size adapt to the requirements of the problem.

In this chapter the notation c_i will keep the same meaning, indicating vectors with norm $|c_i| = 1$ for FHP and HPP and $|c_i| = \sqrt{2}$ for FCHC. However the distance between the lattices vertices will be $h|c_i|$ where h is an adjustable mesh size, and the duration of the time step will be τ. The velocity of the particles will thus be hc_i/τ and all the previous formulae may be recovered in the new scaling using this new definition of the particle velocities. To simplify the discussion, and because this book is almost always concerned with incompressible flow, we do not introduce a dimension of mass.

Thus the mass and momentum invariants are

$$\rho = N_0 + \sum_{i>0} N_i \tag{7.1}$$

and

$$\rho \mathbf{u} = \sum_i N_i \mathbf{c}_i h/\tau. \tag{7.2}$$

The momentum flux tensor is

$$\Pi_{\alpha\beta}^{(P)} = \sum_i N_i^{(P)} c_{i\alpha} c_{i\beta} h^2/\tau^2 \tag{7.3}$$

and the new pseudo-equilibrium is

$$N_i^{(P)} = f \left[1 + \frac{D\tau}{c^2 h} c_{i\alpha} u_\alpha + \left(G_m Q_{i\alpha\beta} + G_c c^2 \delta_{\alpha\beta} \right) \tau^2 u_\alpha u_\beta/h^2 \right] \tag{7.4}$$

and

$$N_0^{(P)} = f(b_c - b_m G_c c^2 \tau^2 u^2/h^2). \tag{7.5}$$

The Boltzmann equation reads

$$N_i(\mathbf{x} + h\mathbf{c}_i, t + \tau) = (1 + \lambda)N_i(\mathbf{x}, t) - \lambda N_i^{(P)}(\mathbf{x}, t). \tag{7.6}$$

Since this only amounts to a change of the units of length and time, the resulting equations are unchanged. We recover equation (6.24) with the kinematic viscosity ν given by

$$\nu = \frac{c^2 h^2}{(D+2)\tau} \left(-\frac{1}{\lambda} - \frac{1}{2} \right). \tag{7.7}$$

When the rescaling is performed, care must be taken to select λ in such a way as to keep ν to the desired value. It is in practice more useful to express λ as a function of the desired viscosity:

$$\lambda = - \left[\frac{1}{2} + \frac{\nu\tau(D+2)}{c^2 h^2} \right]^{-1}. \tag{7.8}$$

The requirement, first stated in Section 2.1, that the hydrodynamic length scale be much larger than the mean free path is now translated into the condition $L_{\text{hydro}} \gg h$. The analysis of the limit $h/L_{\text{hydro}} \to 0$ is a classical topic of the analysis of finite difference schemes, to which we now turn.

7.2   Linearized stability and convergence analysis

The study of numerical stability of lattice-Boltzmann schemes is patterned after similar analyses for finite difference schemes. As in the case of finite difference schemes, stability is important because in unstable schemes the

computed fields diverge exponentially away from the "true" solutions of
the Navier-Stokes equation*

To make the stability problem tractable by elementary methods, the
Boltzmann method will be linearized around uniform values $N_i^{(P)}$ corresponding to a pseudo equilibrium for uniform fields u_0, ρ_0. The behavior
of small perturbations away from the u_0, ρ_0 equilibrium in the lattice-
Boltzmann populations depends on the eigenvalues of a suitably defined
linearized time evolution operator. The scheme is said to be stable if this
linearized operator leads to exponentially growing solutions and unstable
otherwise.

An elementary stability analysis can be performed in the following way.
Let the populations $N(t)$ be a time-dependent, uniform perturbation of
the constant pseudo-equilibrium populations. Then

$$N(t) = N^{(P)} + V\xi^t, \tag{7.9}$$

where V is an eigenvector of Λ and ξ is a constant. Then inserting into
equation (6.7) one obtains

$$(\xi - 1)V = \Lambda V. \tag{7.10}$$

As Λ is symmetric $\lambda = \xi - 1$ is a real eigenvalue. From (7.9), linear
stability·requires that $|\xi| < 1$. Thus any eigenvalue λ of Λ must verify
$-2 < \lambda < 0$. From equation (6.25) this seems to allow all positive values
of the shear viscosity.

The above analysis of stability is however very incomplete. A more
classical analysis of stability, akin to the stability analysis performed for
finite-difference methods, can be made in the following way. Let the
hydrodynamical variables vary as

$$\mathbf{u}(\mathbf{x}, t) = \mathbf{u}_0 + \epsilon \mathbf{u}_1 \exp(i\mathbf{k} \cdot \mathbf{x} + st) \tag{7.11}$$

$$\rho(\mathbf{x}, t) = \rho_0 + \epsilon \rho_1 \exp(i\mathbf{k} \cdot \mathbf{x} + st) \tag{7.12}$$

where $\mathbf{k}$ is a wavevector in the reciprocal lattice of $\mathcal{L}$, s is the growth
rate, and ρ_0 and $\mathbf{u}_0$ are a fixed density and velocity. The computation
will be unstable if the real part $\text{Re}(s)$ is positive. The full analysis is
rather intricate and has been carried out only numerically.

The results are rather complex since they depend on parameters defining the method b_c and λ and on the dimensionless parameters defining
the perturbation. These are kh and the Courant-Friedrichs-Levy (CFL)
number $C_u = \tau u_0/h$. The orientation of the velocity $\mathbf{u}_0/u_0$ also plays a
role. There are two important dimensionless groups to notice, the CFL

---

* Moreover, it is known in finite difference theory that stability and consistency ensures for
  linear problems convergence of the numerical results to the true solutions as the lattice
  spacing h and the time step τ go to zero.

C_u which characterizes the velocity of the flow in lattice variables[†] and $C_\nu = \nu\tau/h^2$, which is the viscosity in lattice variables. The following trends may be extracted from the numerical results:

- The CFL number C_u should remain below a certain upper stability boundary C_m.

- This boundary decreases with decreasing C_ν. For $C_\nu = 0$ (which corresponds to $\lambda = -2$) some estimates yield a finite stability range bounded by a small positive C_m.

- For the hexagonal lattice, maximal stability is achieved when $b_c = 4$, that is when 2/3 of the mass is in the rest particle population.

- For large values of the dimensionless viscosity C_ν there is a finite maximal velocity C_m whose precise value depends on the model considered. For instance, when $b_c = 4$, $C_m \simeq 0.4$.

The above results are specific to the linearized stability analysis for uniform base flows ρ_0, $\mathbf{u}_0$. Because real flows are nonlinear and non-uniform, numerical practice may contain surprises. In many circumstances, small values of ν are known to produce instabilities even at small dimensionless flow velocity C_u. A safer prescription is to keep

$$C_\nu > C_1 C_u^2 \qquad (7.13)$$

where C_1 is a dimensionless number depending on the specific model considered.

It is also interesting to investigate the behavior of the model when the rescaled viscosity $\nu\tau/h^2$ is increased, as may happen when h^2 is left to go to zero faster than τ. This is equivalent to increasing the viscosity. This high viscosity limit may be understood through an analogy with the Boolean case, where the increase of viscosity corresponds to an increase in the mean free path ℓ_{mfp}. The analogy leads to $|\lambda| \simeq 1/\ell_{\mathrm{mfp}}$. For convergence to the Navier-Stokes equation, the hydrodynamic scale L_{hydro} must verify $L_{\mathrm{hydro}} \gg \ell_{\mathrm{mfp}}$. We thus require

$$|\lambda| \gg 1/L_{\mathrm{hydro}}. \qquad (7.14)$$

An interesting empirical fact is that as $\nu\tau/h^2$ is increased the method does not become unstable. This in marked contrast with comparable explicit finite difference methods.

[†] The CFL number is well known in standard finite difference analysis. For explicit finite difference schemes, there is an upper limit for the values of C_u, above which the scheme becomes unstable.

7.3 Forcing the flow

External forces, such as gravity, may be represented by adding a fixed amount of momentum at each site. The general idea is the same as in a Boolean model, but as usual, the lattice-Boltzmann version is simpler. For a force $\mathbf{f}$, the modified equation reads

$$N_i(\mathbf{x} + h\mathbf{c}_i, t + \tau) - N_i(\mathbf{x}, t) = \Delta_i[\mathbf{N}(\mathbf{x}, t)] + \frac{\tau D}{b_m c^2}\mathbf{f} \cdot \mathbf{c}_i. \qquad (7.15)$$

The momentum balance equation then becomes

$$\partial_t \rho u_\alpha = -\partial_\beta \Pi_{\alpha\beta}^{(P)} + f_\alpha. \qquad (7.16)$$

Using the change of variable (6.19), still neglecting compressibility effects, and adding the viscous terms we get

$$\partial_t \mathbf{v} + \mathbf{v} \cdot \nabla \mathbf{v} = -\nabla\omega + \nu\nabla^2\mathbf{v} + \mathbf{f}. \qquad (7.17)$$

For instance, taking $\mathbf{f} = -g\mathbf{E}_z$, with g the gravity acceleration, we obtain the usual Navier-Stokes equation in a gravity field. Pressure gradients may also be represented by a constant forcing. Indeed adding the force $A\mathbf{E}_x$ is equivalent to changing the pressure from ω to $\omega - Ax$.

7.4 Boundary conditions

There are various types of boundary conditions that one may impose on a flow domain. Typically, one may have solid walls (moving with velocity V_S or at rest), flow entry and exit conditions, or free slip conditions on rigid surfaces.

Inflow and outflow boundary conditions may be represented with various degrees of accuracy. Assume that the macroscopic condition is flow entering the lattice at some pressure ω_0. In the simplest case, one defines a series of marginal sites, just outside the fluid domain near entry or exit boundaries. To fix ideas, we consider the hexagonal lattice. There are two particular cases that are represented in Figure 7.1: the boundary may be parallel (line CD), or perpendicular (line AB) to a lattice axis.

In the parallel case, there are two lattice directions cut by the boundary. They correspond to indices $i = 1$ and 2 (Fig. 7.1). At the end of each collision step, the "marginal" sites nearest to the boundary of the domain at $x = x_w$ get new populations $N_1'(\mathbf{x}_w, t)$ and $N_2'(\mathbf{x}_w, t)$. These populations are fixed to $N_i' = N_i^{(P)}(\rho_0, \mathbf{u}_0)$ where the notation $N_i^{(P)}(\rho_0, \mathbf{u}_0)$ indicates that we select the pseudo-equilibrium corresponding to some density $\rho_0 = \omega_0/c_s^2$ and velocity $\mathbf{u}_0$. In this way the populations after

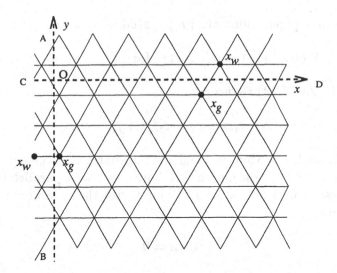

Fig. 7.1. Two simple kinds of inflow and outflow conditions. The domain boundary is either one of the dotted lines. In both cases the first site in the domain is x_g and the nearest site outside is x_w.

propagation on the first site in the gas $\mathbf{x}_g$ are

$$N_i(\mathbf{x}_g) = N_i^{(P)}(\rho_0, \mathbf{u}_0) \quad \text{for} \quad i = 1, 2. \tag{7.18}$$

The velocity $\mathbf{u}_0$ should in principle not be too far from the flow velocity $\mathbf{u}_b$ established inside the flow across the boundary. Indeed from (6.13) we expect that

$$N_i(\mathbf{x}_g) \simeq N_i^{(P)}(\rho_0, \mathbf{u}_b) \quad \text{for} \quad i = 1, 2 \tag{7.19}$$

and thus $\mathbf{u}_b \simeq \mathbf{u}_0$. If condition (7.19) is violated there will be a region in which the populations adapt to the conditions prevailing in the interior of the computational domain. Although the computation may still proceed in these conditions, results will be less accurate.

Boundary conditions on a solid wall may be implemented through a similar procedure. Let us again choose a wall orientation parallel to a lattice orientation. After the streaming step, one keeps track of particles arriving on the wall sites $\mathbf{x}_w$ just as for the other sites

$$N_i(\mathbf{x}_w, t) = N_i'(\mathbf{x}_g, t - \tau) \quad \text{for} \quad i \quad \text{such that} \quad \mathbf{x}_w = \mathbf{x}_g + h\mathbf{c}_i. \tag{7.20}$$

During the collision step, the usual collisions are not performed on the wall sites. Instead, particles on wall sites reverse direction in the so-called bounce-back rule. Choose a labelling of the velocities such that $\mathbf{c}_i = -\mathbf{c}_{i+3}$. Then let the populations after bounce-back be

$$N_i'(\mathbf{x}_w, t) = N_{i+3}(\mathbf{x}_w, t). \tag{7.21}$$

Finally the new populations are propagated back to the gas sites:

$$N_i(\mathbf{x}_g, t + \tau) = N_i'(\mathbf{x}_w, t) \quad \text{for} \quad \mathbf{x}_g = \mathbf{x}_w + h\mathbf{c}_i. \qquad (7.22)$$

Equations (7.20 - 7.22) reduce to

$$N_i(\mathbf{x}_g, t + \tau) = N_{i+3}'(\mathbf{x}_g, t - \tau). \qquad (7.23)$$

Why should such a procedure yield the appropriate boundary condition? To fix ideas, we will consider a flat horizontal wall parallel to the line $y = 0$. A first remark is that no mass may cross the solid boundary. Thus the condition

$$\mathbf{u} \cdot \mathbf{n} = 0 \qquad (7.24)$$

(where $\mathbf{n}$ is the vector normal to the boundary) is realized. This is the appropriate boundary condition on a solid wall for a fluid. However it is not clear exactly *where*, with respect to the lattice axes, this condition is realized. It is likely that somewhere between $\mathbf{x}_w$ and $\mathbf{x}_g$ is a good guess, but we want a more definite answer. The accurate location will be written $y = y_b$.

The full analysis will be performed for viscous fluids. In that case, however, we want to realize the no-slip condition

$$\mathbf{u} = 0 \quad \text{for} \quad y = y_b. \qquad (7.25)$$

The analysis of the bounce-back condition is, just as the derivation of the Navier-Stokes equation, based on an approximate — asymptotic — analysis in the limit of length scales $L_{\text{hydro}} \gg h$ where h is the lattice spacing. In that limit one may represent the field near the boundary by a Taylor expansion near $y = y_b, x = 0$. Using (7.25) the expansion is of the form

$$\begin{aligned} u_x &= \omega(y - y_b) + a_1(y - y_b)^2 + a_2(y - y_b)x + \cdots \\ u_y &= a_0(y - y_b) + b_1(y - y_b)^2 + b_2(y - y_b)x + \cdots \end{aligned} \qquad (7.26)$$

where the dots indicate cubic and higher order terms. Time dependence is omitted as we assume that the flow changes little over one time step. Using (6.12) one finds $a_0 = 0$. Dimensionally we find that a_2 and b_2 are of the order $1/L_{\text{hydro}}$. Thus near the boundary we may neglect x^2, xy and y^2 terms and the flow is the shear flow of equation (4.50). The analysis there may be repeated, replacing the equilibrium populations by the pseudo-equilibrium ones. The populations are given in equation (4.52) which after appropriate substitutions yields[‡]

$$N_i(\mathbf{x}) = N_i^{(P)}(\rho_0, \mathbf{u}(\mathbf{x})) + \frac{2f\omega}{\lambda} Q_{ixy} \qquad (7.27)$$

where $\mathbf{u}(\mathbf{x})$ is the velocity field given in (7.26). As the flow is stationary, equation (7.23) reduces to

$$N_i(\mathbf{x}_g) = N'_{i+3}(\mathbf{x}_g). \qquad (7.28)$$

We take the origin $\mathbf{x} = 0$ at $\mathbf{x}_g$ and replace (7.27) in (7.28) using also (6.3) and (6.9). Then one finds a solution for

$$y_b = -(\sqrt{3}/4)\, h, \qquad (7.29)$$

half-way between the gas and wall sites.

Consider now a wall perpendicular to a lattice direction. Depending on the sites, the cut links are now $i = 1, 5$ and 6 or just $i = 6$. The wall is located at $x = x_b$, and the first order flow is

$$u_x = 0$$
$$u_y = \omega(x - x_b). \qquad (7.30)$$

The same calculation as above yields

$$x_b = -h/4. \qquad (7.31)$$

These are quite important results, since the knowledge of the exact location of the boundary is critical to many flow configurations. Examples are the estimation of the permeability through porous media or the viscosity of suspensions. The simple calculation in this chapter is however only the beginning of a much longer story. When parabolic terms are taken into account (that is, when $\mathcal{O}(y^2)$ terms in equations (7.26) are no longer neglected) the location of the effective wall is different.

7.5 Notes

As for the previous chapter, a lot of additional information may be found in Ref. [6.1]. The stability analysis in this chapter is inspired in a large part by:

7.1 Sterling, J. D. and Chen, S. (1995). Stability analysis of lattice Boltzmann methods, *J. Comput. Phys.* **123**, 196–206.

We have also been inspired by the stability analysis in:

[‡] It is interesting to remark that equation (7.27) yields an exact steady solution of the Boltzmann equation.

7.2 Lafaurie, B. (1995). Modélisation de la convection par un méthode de gaz sur réseau et technique de suivi d'interface, Ph. D. thesis, Université de Paris-Sud (Paris XI Orsay).

The above reference also contains an interesting validation of the method in which the lattice-Boltzmann results were compared to a referenced benchmark test case of large Prandtl number Rayleigh-Bénard convection. The analysis of boundary conditions is inspired by:

7.3 Cornubert, R., d'Humières, D., and Levermore, C. D. (1991). A Knudsen layer theory for lattice gases. *Physica D* **47**, 241-259.

A more refined analysis of boundary conditions may be found in

7.4 Ginzbourg, I. and Adler, P. M. (1994). Boundary flow condition analysis for the three-dimensional lattice-Boltzmann model, *J. Physique II France* **4**, 191-214.

7.5 Ginzbourg, I. and d'Humières, D. (1996). Local second-order boundary method for Lattice Boltzmann models, *J. Stat. Phys.* **84**, 927-971.

7.6 Ginzbourg, I., Giraud, L. and d'Humières, D. (1995). Efficient lattice-Boltzmann calculations in porous media, in preparation.

8

Miscible fluids

In the previous chapters we have restricted our discussion to models of simple fluids such as water. But what if the water contains some dye? Thus we now consider models of fluid mixtures.

In this chapter we consider the simplest mixtures—those composed of two miscible fluids. We will construct models of miscible fluids from straightforward extensions of the models of the previous chapters. To illustrate an application of these miscible-fluid models, we will close this chapter with a summary of a study of *passive scalar dispersion* in a slow flow.

A discussion of miscible lattice gas mixtures is the first step toward an understanding of lattice gas models of *immiscible* fluids, a subject which we shall take up in the following chapter.

8.1   Boolean microdynamics

Miscible lattice-gas mixtures are constructed by adding a second type of particle and letting it evolve passively. This is roughly equivalent to injecting a fluid with a colored dye that allows one to see the fluid motions but does not affect the flow itself.

For convenience, we usually distinguish between particle type by assuming that the particles are colored. Our favorite mixtures are *red* and *blue*. (Aside from a certain aesthetic appeal, this choice also nicely avoids the political pitfalls of *black* and *white*.) In a miscible lattice-gas mixture the color of the particles is a property that is carried with them, but the evolution of the particles is no different than it would be if they were not colored. In other words, when such a model is put on a hexagonal lattice, it evolves just like the FHP model of Figure 1.1, but with one additional complication: the color of the particles is redistributed after collision. Usually color is conserved, and the resulting macroscopic model includes an equation that describes the advection and diffusion of color.

We use $s(\mathbf{x})$ to denote the state at a lattice site located at position $\mathbf{x}$. For a red-blue mixture without rest particles on a hexagonal lattice, s is the Boolean 12-vector

$$s = (r_1, b_1, \ldots, r_6, b_6), \tag{8.1}$$

where, in analogy with our previous developments, $\mathbf{r} = (r_i)_i$ and $\mathbf{b} = (b_i)_i$ are Boolean variables that indicate the presence (1) or absence (0) of a red or blue particle, respectively, moving with unit speed in the direction $\mathbf{c}_i$. Although red and blue particles may be at the same site, for simplicity we do not allow them to move with the same velocity at the same site. Thus r_i and b_i cannot both equal one for the same i, and we have, as before, the Boolean variable

$$n_i = r_i + b_i \tag{8.2}$$

to indicate whether a particle of *either* color is moving in direction $\mathbf{c}_i$.

The microdynamical equations of the mixture are

$$r_i(\mathbf{x} + \mathbf{c}_i, t + 1) = r_i'(\mathbf{x}, t), \qquad b_i(\mathbf{x} + \mathbf{c}_i, t + 1) = b_i'(\mathbf{x}, t), \tag{8.3}$$

where r_i' and b_i' are post-collision, pre-propagation states. Because color plays the role of a passive scalar property in miscible models, these two equations are coupled only through the conservation of color. Thus, as in equation (2.33), we still have

$$n_i'(\mathbf{x}, t) = n_i(\mathbf{x}, t) + \Delta_i[\mathbf{n}(\mathbf{x}, t)] \tag{8.4}$$

and

$$n_i(\mathbf{x} + \mathbf{c}_i, t + 1) = n_i'(\mathbf{x}, t) \tag{8.5}$$

to indicate that the post-collision *uncolored* state $\mathbf{n}'$ does not depend on the color distribution of the pre-collision state $\mathbf{r}, \mathbf{b}$. Instead, any post-collision distribution of color is possible providing that the number of red particles is conserved, i. e.,

$$\sum_i r_i' = \sum_i r_i. \tag{8.6}$$

Indeed, one may conceive of the collision process as having two steps. First, the particles collide, regardless of their color, according to equation (8.4). Then the particles are recolored in any way that satisfies equation (8.6).

In addition to the Boolean mass-balance and momentum-balance equations (2.41) and (2.42), equation (8.6) gives an equation for the balance of red mass,

$$\sum_i r_i(\mathbf{x} + \mathbf{c}_i, t + 1) = \sum_i r_i(\mathbf{x}, t). \tag{8.7}$$

Conservation of mass of course gives a similar—and redundant—equation for the blue particles.

8.2 Convection-diffusion equation

The addition of the conservation of color yields one more macroscopic equation to describe the mixture. To see this, we first define the ensemble averages

$$R_i = \langle r_i \rangle \quad \text{and} \quad B_i = \langle b_i \rangle, \tag{8.8}$$

which represent the average populations of red and blue particles, respectively, while

$$N_i = \langle n_i \rangle = R_i + B_i \tag{8.9}$$

is the same mean particle population we have used previously. Additionally, we define the dimensionless red concentration $\theta = \sum_i R_i / \sum_i N_i$.

By the same arguments that allowed us to derive the continuity equation (2.49) from the Boolean mass-balance equation (2.41), the balance of red mass given by equation (8.7) gives

$$\partial_t(\theta\rho) = -\partial_\alpha(\theta\rho u_\alpha). \tag{8.10}$$

We are not yet done, however, because we must incorporate the physics of diffusion. The situation is roughly analogous to our previous addition of viscous stress to the Euler equation. Our task is considerably easier here, however, because we are dealing with the transport of a scalar quantity— the color—as opposed to the transport of momentum, a vector.

To see what is happening, it is perhaps useful to imagine that red particles are "hot" and blue particles are "cold", and therefore that θ is a kind of dimensionless temperature. Then, we would expect any temperature anomalies to smooth out or *diffuse* due to the random walks taken by each individual particle as it propagates and collides. To account for this smoothing effect, we must incorporate a diffusive current **J** of red particles into the right-hand side of equation (8.10). Since color is conserved, we need the divergence $\partial_\alpha J_\alpha$ and therefore write

$$\partial_t(\theta\rho) = -\partial_\alpha(\theta\rho u_\alpha + J_\alpha). \tag{8.11}$$

It remains to specify the current **J**. In real mixtures, a phenomenological relation known as *Fick's law* usually holds. Fick's law is a linear-response relation that states that diffusive currents are proportional to concentration gradients, provided that the concentration gradients are small. We make precisely the same assumption here. In other words, we assume that the relaxation of the lattice to its equilibrium state may be described by a linear response theory similar to the one we constructed for viscosity in Chapter 4. Thus we write

$$J_\alpha = -D_0 \partial_\alpha (\theta \rho), \tag{8.12}$$

where D_0, a positive constant, is the *diffusion coefficient*. Substituting equation (8.12) into (8.11) we obtain

$$\partial_t (\theta \rho) + \partial_\alpha (\theta \rho u_\alpha) = D_0 \partial^2_{\alpha\alpha} (\theta \rho). \tag{8.13}$$

Finally, using equation (2.49), converting to vector notation, and neglecting any density variations, we obtain

$$\partial_t \theta + \mathbf{u} \cdot \nabla \theta = D_0 \nabla^2 \theta. \tag{8.14}$$

Equation (8.14), a *convection-diffusion* equation, describes the combined effects of the diffusive transport of color due to concentration gradients and the advective transport due to flow. The diffusion coefficient D_0 depends on the precise collision rules, and, as we have already seen for the viscosity, a particular eigenvalue of the linearized collision operator. We turn now to its explicit evaluation.

8.3   Diffusion coefficient

Our task here is simpler than the calculation of viscosity in Chapter 4, because diffusion involves the transport of a scalar rather than a vector quantity. Moreover we have already laid much of the conceptual groundwork for such a procedure. So in the following we only sketch the calculation.

First, we combine the red and blue mean populations into a single population vector $\mathbf{M}$ such that

$$\mathbf{M} = (R_1, B_1, \ldots, R_6, B_6). \tag{8.15}$$

Then, analogous to the Boltzmann equation (4.6), we write

$$M_i(\mathbf{x} + \hat{\mathbf{c}}_i, t + 1) - M_i(\mathbf{x}, t) = \Delta_i[\mathbf{M}(\mathbf{x}, t)], \tag{8.16}$$

where $i = 1, \ldots, 12$ and $\hat{\mathbf{c}}_i$ is the velocity vector of particles indexed by i such that

$$(\hat{\mathbf{c}}_i)_i = (\mathbf{c}_1, \mathbf{c}_1, \ldots, \mathbf{c}_6, \mathbf{c}_6). \tag{8.17}$$

The collision operator Δ_i is a two-color generalization of operators like that given by equation (4.35).

We are interested in calculating the diffusive flux of color that responds to a color gradient. We consider a linear gradient in the red fraction θ such that

$$\theta(y) = \omega y. \tag{8.18}$$

For no net flow of mass and constant θ, the equilibrium red and blue populations are

$$R_i^{eq} = \theta f, \qquad B_i^{eq} = (1 - \theta)f, \qquad i = 1, \ldots, 6, \qquad (8.19)$$

where f is again the reduced density. We expand the populations R_i and B_i for small color gradients $\omega \ll 1$:

$$R_i(\mathbf{x}) = R_i^{eq}(\mathbf{x}; \theta) + \epsilon_i(\mathbf{x}) + \mathcal{O}(\omega^2) \qquad (8.20)$$
$$B_i(\mathbf{x}) = B_i^{eq}(\mathbf{x}; \theta) - \epsilon_i(\mathbf{x}) + \mathcal{O}(\omega^2). \qquad (8.21)$$

Here ϵ_i is an unknown correction of order ω. After substituting this expansion into equation (8.16), we obtain

$$M_i(\mathbf{x} + \hat{\mathbf{c}}_i, t + 1) - M_i(\mathbf{x}, t) = \sum_j \Lambda_{ij} \hat{\epsilon}_j (-1)^{j-1} + \mathcal{O}(\omega^2), \qquad (8.22)$$

where $\hat{\epsilon}_i$ is related to ϵ_i as $\hat{c}_i$ is to c_i, and the linearized collision operator

$$\Lambda_{ij} = \left. \frac{\partial \Delta_i}{\partial M_j} \right|_{R_i = \theta f, \ B_i = (1-\theta)f} \qquad (8.23)$$

As in the developments leading to equation (4.11), we may approximate equation (8.22) as

$$M_i^{eq}(\mathbf{x} + \hat{\mathbf{c}}_i) - M_i^{eq}(\mathbf{x}) = \sum_j \Lambda_{ij} \hat{\epsilon}_j (-1)^{j-1}, \qquad (8.24)$$

where we have kept only terms of order ω. Substitution of the color gradient (8.18) and the equilibrium population (8.19) then yields

$$f \omega \hat{c}_{iy} (-1)^{i-1} = \sum_j \Lambda_{ij} \hat{\epsilon}_j (-1)^{j-1}. \qquad (8.25)$$

Investigation of the 12×12 linearized collision operator Λ_{ij} in the manner of Chapter 4 shows that it has an eigenvector

$$\mathbf{v} = (c_{1y}, -c_{1y}, \ldots, c_{6y}, -c_{6y})^T \qquad (8.26)$$

with a non-zero eigenvalue λ_D. We thus have

$$\epsilon_i = f \omega c_{iy} / \lambda_D. \qquad (8.27)$$

For the concentration gradient (8.18), we have

$$J_y = -D_0 \partial_y (\theta \overline{\rho}) = -D_0 \overline{\rho} \omega = -12 f \omega D_0 / \sqrt{3}, \qquad (8.28)$$

where $\overline{\rho}$ is the mass per unit area as defined by equation (4.59) and the first equality comes from an assumption of a Fickian flux of the form (8.12). Taking the y-axis midway between $\mathbf{c}_1$ and $\mathbf{c}_2$ as in Figure 4.2, we also have

$$J_y = R_1(\mathbf{x} + \mathbf{c}_1) + R_2(\mathbf{x} + \mathbf{c}_2) - R_4(\mathbf{x}) - R_5(\mathbf{x}), \qquad (8.29)$$

which is the red flux passing through a line of unit length positioned parallel to c_6 and between $\mathbf{x}$ and $\mathbf{x} + c_1$. Using equations (8.18), (8.19), (8.20), and (8.27) to calculate the right-hand side of equation (8.29), we find, after substitution into equation (8.28),

$$D_0 = -\frac{1}{2}\left(\frac{1}{\lambda_D} + \frac{1}{2}\right). \tag{8.30}$$

The diffusion coefficient may of course also be measured from a simulation. There are a number of ways to do this. Perhaps the simplest method is to initialize a periodic lattice with a concentration wave

$$\theta(\mathbf{x}, t = 0) = 0.5 + \epsilon\cos(\mathbf{k}\cdot\mathbf{x}) \tag{8.31}$$

for a specific wavevector $\mathbf{k}$ and $\epsilon \ll 1$. Solving equation (8.14) with this initial condition gives the solution

$$\theta(\mathbf{x}, t) = \theta(\mathbf{x}, t = 0)e^{-D_0 k^2 t}. \tag{8.32}$$

The diffusion coefficient D_0 may then be inferred from the exponential decay of the concentration wave.

Alternatively one may fix boundaries at $y = 0$ and $y = L$ to have concentrations $\theta = 0.5 + \omega L/2$ and $\theta = 0.5 - \omega L/2$, respectively. Equation (8.29) may then be used to calculate the diffusive flux J_y. Equation (8.12) then gives $D_0 = -J_y/\omega$.

8.4   Lattice-Boltzmann models

A lattice-Boltzmann model for miscible fluids may be constructed within the framework of the relaxation Boltzmann method described in Chapter 6. This model is given by evolution equations for the populations N_i and the relative concentration of one of the constituent species. The evolution of N_i is once again given by equation (6.8), which we repeat here for convenience:

$$N_i(\mathbf{x} + c_i, t + 1) = (1 + \lambda)N_i(\mathbf{x}, t) - \lambda N_i^{(P)}(\mathbf{x}, t). \tag{8.33}$$

The pseudo-equilibrium distribution $N_i^{(P)}$ on a D-dimensional lattice is given by equations (6.3) and (6.4), with parameters b_c, G_m, and G_c as discussed in Chapter 6. The relaxation parameter λ is once again related to the kinematic viscosity ν (of both fluids) by equation (6.25).

We again assume two constituent populations R_i and B_i and the sum rule $N_i = R_i + B_i$. We need only write an equation for the concentration of one of the species. We choose red and define the *red concentration* $\theta_i = R_i/N_i$. The update equation for θ_i reads just like equation (8.33), but with a new relaxation parameter λ_D:

$$\theta_i(\mathbf{x} + \mathbf{c}_i, t+1) = (1 + \lambda_D)\theta_i(\mathbf{x}, t) - \lambda_D \theta_i^{(P)}(\mathbf{x}, t). \qquad (8.34)$$

The pseudo-equilibrium distributions $\theta_i^{(P)}$ are

$$\theta_i^{(P)} = \theta \left(1 + \frac{c_{i\alpha} u_\alpha}{c_s^2} \right) \qquad (8.35)$$

for $i > 0$ and

$$\theta_0^{(P)} = \theta b_c \qquad (8.36)$$

for the red mass at rest. Here $\theta = \sum_i R_i / \sum_i N_i$ is the average concentration at a site and the soundspeed $c_s^2 = b_m c^2/(bD)$. One may verify that equations (8.34), (8.35), and (8.36) conserve the red mass density $\theta\rho$.

Since the evolution of N_i is precisely as in Chapter 6, incompressible hydrodynamic equations for mass density ρ and momentum density $\rho\mathbf{u}$ may be obtained as described previously. These equations are then coupled to the evolution of color. Analogous to our derivation of the mass-balance equation (2.44), the balance of color gives

$$\partial_t \sum_i \theta_i = -\partial_\alpha \left(\frac{1}{b} \sum_i \theta_i^{(P)} c_{i\alpha} \right). \qquad (8.37)$$

Substitution of equation (8.35) and evaluation of the summations then gives

$$\partial_t \theta = -\partial_\alpha(\theta u_\alpha). \qquad (8.38)$$

The isotropy of second order tensors (Property 5.4) ensures that the diffusive term has the correct form. Converting to vector notation and assuming incompressibility, we obtain

$$\partial_t \theta + \mathbf{u} \cdot \nabla \theta = D_0 \nabla^2 \theta. \qquad (8.39)$$

The diffusion coefficient D_0 may be calculated as in the previous section. The result in D dimensions is

$$D_0 = -c_s^2 \left(\frac{1}{\lambda_D} + \frac{1}{2} \right). \qquad (8.40)$$

It is interesting to note that the equilibrium distribution (8.35) and (8.36) for the color fraction contains only terms up to first order in u, whereas $N_i^{(P)}$ contains, as always, terms up to second order. The reason is simple. To obtain the convective term $(\mathbf{u} \cdot \nabla)\mathbf{u}$ in the Navier-Stokes equation, second order terms in $N_i^{(P)}$ are required. (Otherwise one obtains the time-dependent Stokes equation (6.27).) The convection-diffusion equation (8.39), however, contains no terms of second order in velocity. The same reasoning may be used to show that diffusive behavior alone requires

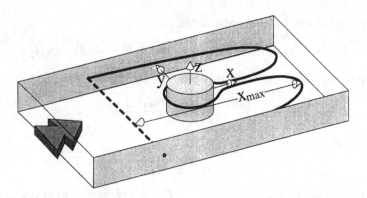

Fig. 8.1. Experimental setup for an experiment on passive-tracer dispersion [8.16]. Two parallel glass plates are closely spaced, and a cylinder is placed upright in the center of the cell. The fluid is forced as shown and a tracer fluid, indicated by the dashed line, is injected into the flow at time $t = 0$. At time t_0 the tracer extends a distance X_{max}.

only isotropy of second-rank velocity-moment tensors, whereas we saw in Chapter 4 that isotropic hydrodynamics requires isotropy of fourth-rank tensors. Thus two-dimensional diffusion in the absence of flow may be simulated on a square lattice.

8.5   Passive tracer dispersion

Whereas a passive tracer in a stationary fluid simply diffuses, the combined effects of diffusion and velocity gradients in a moving fluid cause the tracer to be *dispersed*. Below, we summarize a study by Flekkøy, Oxaal, Feder, and Jøssang [8.15] in which both experiments and lattice-Boltzmann simulations were used to investigate passive tracer dispersion.

Consider the experimental setup of Figure 8.1. Here, two glass plates are separated by a distance which is small compared to the length and the width of the plates. A small cylinder is placed upright between the two plates, and fluid is forced slowly past the cylinder. At time $t = 0$, a tracer fluid, indicated by the dashed line in Figure 8.1, is injected into the flow. The dye—i.e., the passive tracer line—then moves with the flow, and is eventually wrapped partially around the cylinder when it reaches a distance X_{max} at time $t = t_0$. Now suppose the flow is precisely reversed at $t = t_0$. What is the shape of the passive tracer line at time $t = 2t_0$? Although the flow is three dimensional, we may consider it to be two dimensional by an appropriate definition of a forcing term that accounts for viscous drag from the two plates [8.15]. To illustrate what is going on, in Figure 8.2 we show the streamlines of the flow. Because the flow

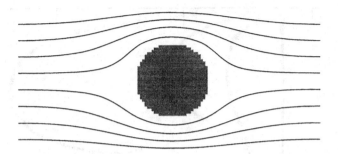

Fig. 8.2. Streamlines for two-dimensional flow past a cylinder, obtained from a lattice-Boltzmann simulation. The flow satisfies the Stokes equations (8.41) and (8.42). Note that the streamlines are symmetric with respect to the cylinder.

is very slow and steady, we may approximate it by neglecting the time derivative and nonlinear term of the Navier-Stokes equations. We are left with the *Stokes equations*

$$\nabla \cdot \mathbf{u} = 0 \tag{8.41}$$

and

$$\nabla p = \mu \nabla^2 \mathbf{u}. \tag{8.42}$$

Notice that flow governed by the Stokes equations is invariant under the transformation $\mathbf{u} \to -\mathbf{u}$, $\nabla p \to -\nabla p$, or, in other words, it is *reversible*. Therefore we expect the same set of streamlines for flow past a cylinder regardless of whether the flow is forced from left to right or from right to left. This is why the streamlines of Figure 8.2 are symmetric with respect to the cylinder.

The reversibility of Stokes flow has an important consequence. Suppose the passive tracer line were perfectly non-diffusive. Then if the flow were reversed (i.e., $\mathbf{u} \to -\mathbf{u}$) at any time t_0 after the tracer was injected, one would expect to recover precisely the initial passive tracer line at time $2t_0$. On the other hand, if there were just a small amount of diffusion, one would naively expect to see a line which has been diffusively broadened, i. e., a line slightly thicker than the original.

How do such expectations compare to experiment? Figure 8.3 shows the evolution of the passive tracer line for flow past the cylinder. The tracer line furthest to the right is shown at time $t \approx t_0$. At this time, just before the flow is reversed, the tracer line is stretched past the cylinder (which is not visible). The three other images of the tracer are shown at later times, with time increasing from right to left. The leftmost image of the tracer is shown at time $t = 2t_0$. A magnification of the central part of this final tracer line is shown in Figure 8.4. For comparison, lattice-

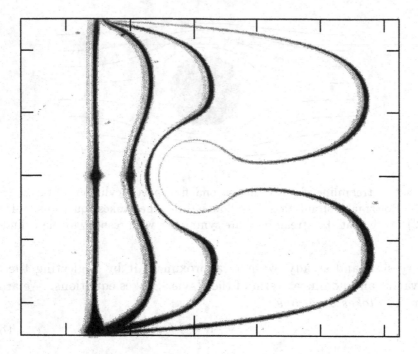

Fig. 8.3. Four snapshots of the return flow of the tracer line in the passive-tracer dispersion experiment described in Figure 8.1 [8.15]. The rightmost tracer line is shown wrapped around the cylinder (which is invisible) at time t_0. Time increases from right to left; the leftmost tracer line is shown at time $2t_0$. Note the broadening of the central portion of the tracer line upon return. The gridmark spacing is 1 cm.

Boltzmann simulations of the experiment are shown in Figure 8.5. Both the experiment and the simulation show the same basic effect: by the time the tracer line has returned to its original position at time $t = 2t_0$, it has undergone significant broadening in the central portion, but nearly no broadening elsewhere. Why is this so?

Notice that, particularly in the simulations shown in Figure 8.5, the tracer line has returned with a cusp-like shape in the center, indicating that the closer the tracer line was to approaching the cylinder directly towards its center, the greater was the dispersion of the tracer. This abrupt increase in dispersion is due to the presence of a *stagnation point* in front of the cylinder, i. e., a point at which there is no fluid motion. To see this, consider the time $T(\mathbf{x})$ that is required for a tracer particle to arrive at a point $\mathbf{x}$. As $\mathbf{x}$ approaches a stagnation point, $T(\mathbf{x}) \to \infty$. In other words, near a stagnation point, small changes in $\mathbf{x}$ can cause large changes in the arrival time T. If there were no diffusion whatsoever, there would be no broadening when the tracer line returns. However a

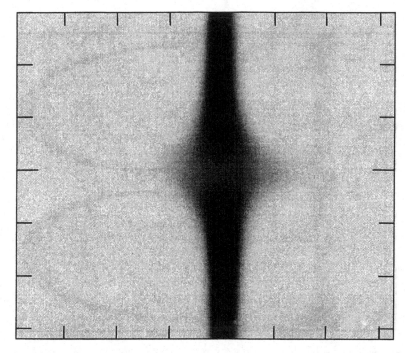

Fig. 8.4. A blowup of the leftmost experimental tracer line in Figure 8.3 [8.15]. The gridmark spacing is now 0.1 cm.

very small amount of diffusion can cause a tracer particle to hop to a streamline closer to the stagnation point on its approach to the cylinder, and therefore lag behind the main tracer line when the flow is reversed. Thus we have the broadening of the tracer in front of the cylinder.

This effect may be quantified. In the absence of flow, the convection-diffusion equation (8.14) reduces to the simple diffusion equation

$$\partial_t \theta(x, y, t) = D_0 \partial^2_{xx} \theta(x, y, t), \tag{8.43}$$

where we have neglected variations in the y-direction. For an arbitrary initial condition $\theta(x, y, 0)$, this equation has the solution

$$\theta(x, y, t) = \frac{1}{\sqrt{4\pi D_0 t}} \int_{-\infty}^{\infty} dx' \theta(x', y, 0) \exp\left(-\frac{(x - x')^2}{4D_0 t}\right). \tag{8.44}$$

The mean-square width of the concentration in the x-direction about a point x_0 at any time t is

$$\sigma_y^2(t) = \frac{1}{\Theta} \int_{-\infty}^{\infty} dx(x - x_0)^2 \theta(x, y, t), \tag{8.45}$$

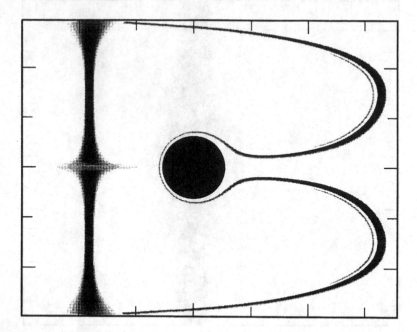

Fig. 8.5. Lattice-Boltzmann simulations of the experiment pictured in Figures 8.3 and 8.4 [8.15]. The tracer line is now shown only at times t_0 and $2t_0$.

where $\Theta = \int_{-\infty}^{\infty} \theta(x, y, 0) dx$. Substitution of equation (8.44) then yields

$$\sigma_y^2(t) = \frac{1}{\Theta} \int_{-\infty}^{\infty} \int_{-\infty}^{\infty} dx dx' (x - x_0)^2 \frac{1}{\sqrt{4\pi D_0 t}} \exp\left(-\frac{(x - x')^2}{4 D_0 t}\right) \theta(x', y, 0).$$

(8.46)

By making the change of variable $x \to x + x'$, we obtain

$$\sigma_y^2 = \frac{1}{\Theta} \int_{-\infty}^{\infty} \int_{-\infty}^{\infty} dx dx' (x + x' - x_0)^2 \frac{1}{\sqrt{4\pi D_0 t}} \exp\left(-\frac{x^2}{4 D_0 t}\right) \theta(x', y, 0).$$

(8.47)

Since the exponential function is even in x, when it is multiplied by odd orders of x the integral over x vanishes and we obtain

$$\sigma_y^2(t) = \frac{1}{\Theta} \left(2 D_0 t \int_{-\infty}^{\infty} dx' \theta(x', y, 0) + \int_{-\infty}^{\infty} dx' (x' - x_0)^2 \theta(x', y, 0)\right),$$

(8.48)

which yields the simple result

$$\sigma_y^2(t) = 2 D_0 t + \sigma_y^2(t = 0).$$ (8.49)

We may now quantify the dispersion. In the absence of flow, the change $\Delta \sigma_y^2 = \sigma_y^2(t) - \sigma_y^2(t = 0)$ would reflect only the diffusive smearing given

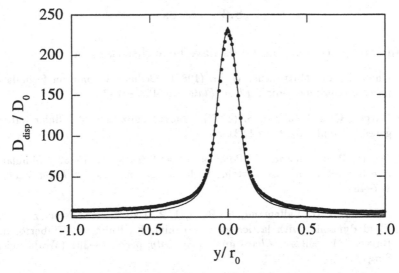

Fig. 8.6. Plot of D_{disp}/D_0 as a function of position across the channel, computed from the returned tracer lines in the experiment (dots) shown in Figure 8.4 and the simulation (full line) shown in Figure 8.5 [8.15]. The normalization factor r_0 is the radius of the cylinder.

by $2D_0 t$. To describe dispersion—the combined effects of diffusion and velocity gradients—we define the effective diffusion coefficient

$$D_{\text{disp}} = \frac{\Delta \sigma_y^2}{2t}. \tag{8.50}$$

In any non-uniform flow we expect $D_{\text{disp}} \neq D_0$.

Figure 8.6 shows a plot of D_{disp}/D_0 as a function of position along the returned tracer line for both the experiment and the simulation. There are two points to be made here. First, we see that the effects of dispersion have caused the effective diffusivity to be enhanced by more than a factor of 200 as the tracer line directly approached the cylinder. Second, the agreement between simulation and experiment is good, especially so in the region of significant dispersion. Thus we may view this result both as a validation of the experimental results by simulation and as a validation of the lattice-Boltzmann method itself. This unusually good agreement led to the suggestion that both experiments and simulations may be used together to estimate very small diffusion coefficients, i.e., by inferring D_0 in a real fluid from matching the ratios D_{disp}/D_0 in analogous simulations and experiments.

8.6   Notes

Boolean models of miscible fluids have been discussed in:

8.1 Chen, H. and Matthaeus, W. H. (1987). Cellular automaton formulation of passive scalar dynamics. *Phys. Fluids* **30**, 1235–1237.

8.2 Burges, C. and Zaleski, S. (1987). Buoyant mixtures of cellular automaton gases. *Complex Systems* **1**, 31–50.

8.3 Clavin, P., Lallemand, P., Pomeau, Y., and Searby, G. (1988). Simulation of free boundaries in flow systems by lattice-gas models. *J. Fluid Mech.* **188**, 437–464.

8.4 d'Humières, D., Lallemand, P., Boon, J. P., Dab, D., and Noullez, A. (1988). Fluid dynamics with lattice gases. In Livi, R., Ruffo, S., Ciliberto, S., and Buiatti, M., editors, *Chaos and Complexity*, page 278–301 (World Scientific, Singapore).

8.5 Baudet, C., Hulin, J. P., Lallemand, P., and d'Humières, D. (1989). Lattice-gas automata: a model for the simulation of dispersion phenomena. *Phys. Fluids A* **1**, 507–512.

8.6 McNamara, G. R. (1990). Diffusion in a lattice gas automaton. *Europhys. Lett.* **12**, 329–334.

Each of these papers covers different aspects of diffusion in hydrodynamic lattice gases. Burges and one of us, for example, consider the passive scalar to be heat, and describe a model of thermal convection. Clavin et al. consider models of chemically reacting mixtures. D'Humières et al. describe collision rules that minimize the diffusion coefficient. This can be useful if Boolean models are used for dispersion studies, as was proposed by Baudet et al. McNamara's paper studies diffusion itself—he shows that the diffusion coefficient grows logarithmically with system size in 2D lattice gases, as is also predicted for 2D fluids in general.

Simpler lattice gases that do not conserve momentum may also be constructed to study diffusion. This is a large field in and of itself that predates momentum-conserving lattice-gas automata. A paper that considers aspects of both fields is:

8.7 Kong, X. P. and Cohen, E. G. D. (1991). A kinetic theorist's look at lattice gas cellular automata. *Physica D* **47**, 9–18.

Non-momentum conserving lattice gases have also been extensively studied in the context of reaction-diffusion systems. Some papers on this subject are:

8.8 Dab, D., Lawniczak, A., Boon, J. P., and Kapral, R. (1990). Cellular automaton model for reactive systems. *Phys. Rev. Lett.* **64**, 2462–2465.

8.9 Dab, D., Boon, J. P., and Li, Y. (1991). Lattice-gas automata for coupled reaction-diffusion equations. *Phys. Rev. Lett.* **66**, 2535–2538.

8.10 Kapral, R., Lawniczak, A., and Masiar, P. (1991). Oscillations and waves in a reactive lattice-gas automaton. *Phys. Rev. Lett.* **66**, 2539–2542.

8.11 Lawniczak, A., Dab, D., Kapral, R., and Boon, J. P. (1991). Reactive lattice-gas automata. *Physica D* **47**, 132–158.

8.12 Boon, J. P., Dab, D., Kapral, R., and Lawniczak, A. (1996). Lattice gas automata for reactive systems. *Phys. Rep.* **273**, 55–147.

Lattice-Boltzmann models of miscible fluids are discussed in:

8.13 Holme, R. and Rothman, D. H. (1992). Lattice-gas and lattice-Boltzmann models of miscible fluids. *J. Stat. Phys.* **68**, 409–430.

8.14 Flekkøy, E. G. (1993). Lattice Bhatnagar-Gross-Krook models for miscible fluids. *Phys. Rev. E* **47**, 4247–4257.

The lattice-Boltzmann method described in this chapter more closely represents the one described in the paper by Flekkøy. Flekkøy's method was also the one used to produce the simulation data of Figures 8.5 and 8.6. These lattice-Boltzmann simulations were part of an experimental and numerical study of hydrodynamic dispersion described in:

8.15 Flekkøy, E. G., Oxaal, U., Feder, J., and Jøssang, T. (1995). Hydrodynamic dispersion at stagnation points: simulations and experiments. *Phys. Rev. E* **52**, 4952–4962.

Further studies of the same nature, including comparisons of the lattice-Boltzmann simulations to simulations performed by a finite-difference method, are reported in:

8.16 Flekkøy, E. G., Rage, T., Oxaal, U., and Feder, J. (1996). Hydrodynamic irreversibility in creeping flow. *Phys. Rev. Lett.* **77**, 4170–4173.

9
Immiscible lattice gases

Our discussion of lattice-gas models now takes a qualitative turn. We continue to study fluid mixtures as in the previous chapter, but now they will exhibit some surprising behavior—they won't like to mix!

This change in direction also steers us towards the heart of this book: models for complex hydrodynamics. The particular kind of complexity we introduce in this chapter relates to interfaces in immiscible fluids such as one might find in a mixture of oil and water. We are all familiar with the kind of bubbly complexity that that can entail. So it seems all the more remarkable that only a revised set of collision rules are needed to simulate it with lattice gases. Indeed, the models of immiscible fluids that we shall introduce are so close to the models of the previous chapters that we call them *immiscible lattice gases*.

This chapter, an introduction to immiscible lattice-gas mixtures, is limited to a discussion of two-dimensional models. In the next chapter, we introduce a lattice-Boltzmann method that is the "Boltzmann equivalent" of the immiscible lattice gas. That then sets the stage for our discussion of three-dimensional immiscible lattice gases in Chapter 11.

9.1   Color-dependent collisions

In the miscible lattice gases of the previous chapter, the collision rules were independent of color. The diffusive behavior derived instead from the redistribution of color after generic colorblind collisions were performed. Aside from some diffusion, the color simply went with the flow.

In immiscible lattice gases (ILG's) color still moves with the flow but it also influences where the flow goes. Specifically, the way in which collisions scatter momentum depends on the distribution of color at neighboring lattice sites.

Figure 9.1 illustrates the ILG microdynamics. To see how the ILG

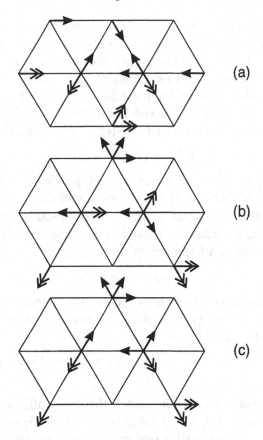

Fig. 9.1. Microdynamics of the immiscible lattice gas, in which the initial condition (a), the propagation step (b), and collision step (c) are displayed as in Figure 1.1. The initial condition and propagation step are the same as before, except that now some particles are red (bold arrows) while others are blue (double arrows). In the collision step, the particles are re-arranged so that, as much as possible, the flux of color is in the direction of the local gradient of color. Compare the middle row here with Figure 1.1 to see how ILG collisions can create a "colorblind" microdynamics different from that created by the single-phase collisions.

compares to the simplest lattice gas, this figure should be compared to Figure 1.1.

The initial state of Figure 9.1 is the same as in Figure 1.1, except that in the ILG some of the particles are red while others are blue. The hopping step, Figure 9.1b, is precisely as before: particles propagate to the neighboring site in the direction of their velocity. The collision step in Figure 9.1c, however, is different. Roughly speaking, the ILG collision rule changes the configuration of particles so that, as much as possible,

red particles are directed towards neighbors containing red particles, and
blue particles are directed towards neighbors containing blue particles.
The total mass, the total momentum, and the number of red (or blue)
particles are conserved. Two examples of this collision rule are seen by
comparing the middle row of Figure 9.1b with that of Figure 9.1c.

More formally, we may define the two-dimensional ILG as follows. As
we did with miscible mixtures in Chapter 8, we define the colored state

$$s = (\mathbf{r}, \mathbf{b}) = (r_0, b_0, \ldots, r_6, b_6). \tag{9.1}$$

Here the bits with indices 0 represent rest particles, which we have included for increased freedom in the choice of collisions, while the bits with
higher indices refer once again to the lattice directions $\mathbf{c}_i$ given by equation (2.34). We again allow at most one particle of either color to move
with a given velocity, and thus have

$$n_i = r_i + b_i \tag{9.2}$$

to indicate the presence of either a red or blue particle moving with velocity $\mathbf{c}_i, i > 0$, or at rest ($i = 0$). We also have the same microdynamical
equations as for miscible fluids:

$$r_i(\mathbf{x} + \mathbf{c}_i, t + 1) = r_i'(\mathbf{x}, t), \qquad b_i(\mathbf{x} + \mathbf{c}_i, t + 1) = b_i'(\mathbf{x}, t), \tag{9.3}$$

but now with $i = 0, \ldots, 6$.

Unlike the case of the miscible model, the post-collision states $\mathbf{r}'$ and $\mathbf{b}'$
depend on the configurations at neighboring sites. Specifically, the post-
collision states are chosen with equal probability among the configurations
$\mathbf{r}'$ and $\mathbf{b}'$ that maximize

$$\mathbf{q}(\mathbf{r}', \mathbf{b}') \cdot \mathbf{f}, \tag{9.4}$$

where

$$\mathbf{q}[\mathbf{r}(\mathbf{x}), \mathbf{b}(\mathbf{x})] = \sum_{i=1}^{6} \mathbf{c}_i [r_i(\mathbf{x}) - b_i(\mathbf{x})] \tag{9.5}$$

is the *color flux*,

$$\mathbf{f}(\mathbf{x}) = \sum_i \mathbf{c}_i \sum_j [r_j(\mathbf{x} + \mathbf{c}_i) - b_j(\mathbf{x} + \mathbf{c}_i)] \tag{9.6}$$

is the *color gradient*, or *color field*, and the set of admissible post-collision
states conserve colored mass and uncolored momentum, i.e.,

$$\sum_i r_i' = \sum_i r_i, \qquad \sum_i b_i' = \sum_i b_i, \tag{9.7}$$

and

$$\sum_i \mathbf{c}_i(r_i' + b_i') = \sum_i \mathbf{c}_i(r_i + b_i). \tag{9.8}$$

How may we interpret such a microdynamics? The effect of the maximization of expression (9.4) is to favor, as much as possible, the uphill diffusion of color. Thus the mixed state may be anti-diffusive and the two colors may spontaneously separate. Once the two colors are separated, the maximization of expression (9.4) results in collision rules that are identical to those of the random-collision seven-bit rest-particle model of Section 3.6. (If you are unsure why, note from equation (9.5) that conservation of momentum allows only one value for $\mathbf{q}$ if either all the r_i's or all the b_i's are equal to zero.) Thus in separated phases we recover the hydrodynamic description of the 7-bit model given in Section 3.7.

We return to these points, and the related issue of surface tension, below. Before doing so, however, we must define the ILG collisions more explicitly. In this way we will not only be better prepared for the theoretical analyses that will follow, but we shall also make clear the algorithmic construction of the model.

ILG collisions are somewhat more complicated than those we have discussed so far because they depend on the color gradient $\mathbf{f}$. Fortunately for us, however, the additional complications are not quite as severe as one might first naively imagine. One simplification arises from the fact that $\mathbf{f}$ does not vary with $\mathbf{r}'$ and $\mathbf{b}'$ in the maximization of expression (9.4); thus only the direction but not the magnitude of $\mathbf{f}$ must be known. In other words, the unit vector $\hat{\mathbf{f}} = \mathbf{f}/|\mathbf{f}|$ suffices. We may simplify things even further, however.

Note that to obtain $\hat{\mathbf{f}}$, we must first obtain the color information at neighboring sites. We define the relative color density at the nearest neighbor site in direction $\mathbf{c}_i$, $i = 1, \ldots, 6$, by

$$\phi_i = \sum_{j=0}^{6}[r_j(\mathbf{x} + \mathbf{c}_i) - b_j(\mathbf{x} + \mathbf{c}_i)] \tag{9.9}$$

$$= \mathcal{M}(s(\mathbf{x} + \mathbf{c}_i)). \tag{9.10}$$

Here a color-counting operator (or, in terms of a computer program, a look-up table) $\mathcal{M}$ has been implicitly defined. Since $\phi_i \in \{-7, -6, \ldots, 7\}$, there are 15^6 possible neighboring color distributions $(\phi_i)_{i=1,\ldots,6}$ that must each be associated with a specific $\hat{\mathbf{f}}$. Obviously there may be many different color distributions $(\phi_i)_i$ that each give the same $\hat{\mathbf{f}}$. Moreover we are not obliged to account for every possible $\hat{\mathbf{f}}$, since small differences in $\hat{\mathbf{f}}$ will have either negligible or no effect on the maximization of (9.4) and the resulting particle dynamics. Thus we may instead substitute an integer *angle code* f_* for $\hat{\mathbf{f}}$. Usually we allow for 36 values of f_*, uniformly distributed from 0 to 2π, plus an additional state to allow for the case $\mathbf{f} = 0$. The transformation of the color distribution $(\phi_i)_{i=1,\ldots,6}$ to the discrete angle code f_* is then symbolically represented by the operator $\mathcal{T}$

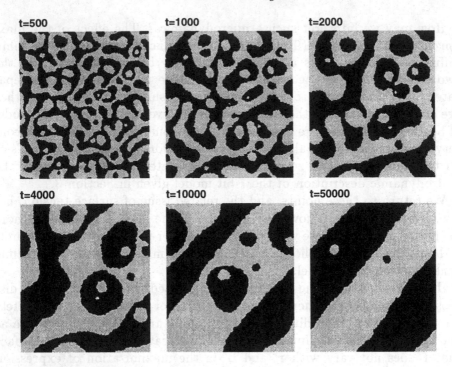

Fig. 9.2. ˙ Phase separation in the immiscible lattice gas [9.6]. The initial condition was a homogeneous random mixture, with 50% red (black) particles, and 50% blue (grey). Time t is given in time steps. Boundaries are periodic in both directions.

such that

$$f_* = \mathcal{T}(\phi_1, \ldots, \phi_6). \tag{9.11}$$

The formal expression for the ILG collisions is then

$$r'_i = \mathcal{C}^r_i(s(\mathbf{x}, t), f_*), \qquad b'_i = \mathcal{C}^b_i(s(\mathbf{x}, t), f_*), \tag{9.12}$$

where the collision operator $\mathcal{C}^j_i$ takes as input the 14-bit state $s(\mathbf{x})$ and the angle code f_*, and gives as output the state of the ith element of species $j \in \{r, b\}$ after collisions have occurred. This description of the microdynamics is then completed by the microdynamical equations (9.3).

9.2 Phase separation

Let's imagine that you, the reader, have constructed an ILG model on your computer. What would you then see?

The most striking feature of the model is its behavior out of equilibrium, as may be seen in Figure 9.2. Here a 256×256 lattice is initialized as a random mixture with an average density $\rho = 4.9$ particles per site, with 50% of the particles red and 50% blue. No forces are applied and there is no net momentum in the system. Moreover there are no walls— the boundaries are periodic in both directions, so that, for example, the nearest neighbors above the top row of the lattice are found on the bottom row. As time progresses, we see that the domains of red and blue grow larger, eventually resulting in a steady state in which one thick blue stripe is parallel to an equally thick red stripe. (The stripe is tilted as a result of the periodic boundary conditions, rather than as a result of any anisotropy in the model.) If you were to examine, say, the red-rich phase, you would find that it would be virtually ($\gg$99%) pure red (and, by symmetry, the same would hold for blue). You would also find that the interfaces between the red and blue phases are sharp, about 1–2 lattice units thick.

Detailed studies show that ILG phase separation is consistent with theoretical models and experimental observations of phase separation in two-component fluids. Rather than considering these issues now, we delay their discussion until Chapter 16 and turn instead to a brief look at the interfaces that are created by the phase separation.

9.3   Surface tension

Among the many interfacial properties worth considering, probably the most important is *surface tension*. Surface tension is defined as a force per unit length, or energy per unit area. It acts to keep interfaces smooth, in a way similar to how a rubber balloon assumes an approximately spherical shape when it is inflated with air.

So that we may be more precise, consider the schematic diagram of a fluid interface in Figure 9.3. Here we have a one-dimensional interface in a two-dimensional fluid. We construct a control volume $ABCD$ around a small piece of the interface with radius of curvature R. The interface subtends an angle 2θ. For real fluids, we expect that all the forces acting on the control volume $ABCD$ will balance. These forces include the pressure forces p_1 and p_2 acting on the concave side AD and convex side BC, respectively, of the control volume. We assume that the control volume is very thin, so we may neglect all forces acting on the sides AB and CD except for capillary forces $\mathbf{f}_c = \sigma \mathbf{n}$, where $\mathbf{n}$ is a unit vector normal to the surface of the control volume. Figure 9.3 shows that the capillary forces $\mathbf{f}_c$ are directed outward so as to maintain tension on the interface. Whereas the horizontal components of the two $\mathbf{f}_c$'s cancel, the vertical components $\sigma \sin \theta$ must be balanced against the vertical component of

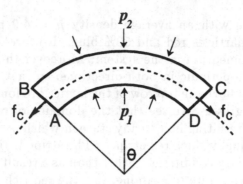

Fig. 9.3. Control volume $ABCD$ (bold curves) around an interface (dashed line). The interface subtends an angle 2θ and its radius of curvature is R. The capillary forces $\mathbf{f}_c$ and the pressures p_1 and p_2 act on the control volume as shown.

the net pressure force integrated over the surface. We thus have

$$\int_{-\theta}^{\theta} (p_2 - p_1) \cos \phi \, Rd\phi + 2\sigma \sin \theta = 0, \qquad (9.13)$$

from which we obtain

$$p_1 - p_2 = \sigma/R. \qquad (9.14)$$

Equation (9.14) is *Laplace's law* for interfaces in two dimensions. We see that flat interfaces separate regions of equal pressure, while curved interfaces create a pressure difference—just as the air inside a balloon is at higher pressure relative to the air outside.

What does all this have to do with ILG interfaces? Proceeding empirically, we may simply initialize the ILG as a red bubble of radius R in a blue "sea" of linear dimension $L \gg R$, and measure the pressure difference between the red bubble and the blue sea. The pressure at any point in either the red or blue phase may be computed from the equation of state (3.57) for a lattice gas with one rest particle. For the case of no net momentum, this gives

$$p = 3\rho/7. \qquad (9.15)$$

Figure 9.4 shows the results of measurements of the pressure difference $\Delta p = p_1 - p_2$ for bubbles ranging from size $R = 4$ to $R = 64$ lattice units. The data are plotted as a function of $1/R$, and we find an excellent empirical agreement with the relation $\Delta p \propto 1/R$. The constant of proportionality, or the slope of the straight line, is the surface tension coefficient σ.

Why do the ILG's microdynamics give rise to Laplace's law? Some insight may be gained from Figure 9.5. Here we have a head-on collision

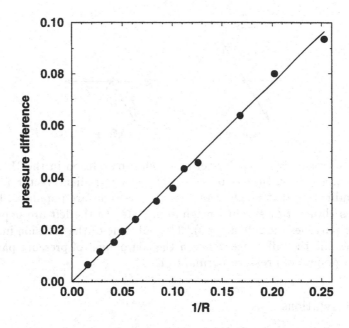

Fig. 9.4. Verification of Laplace's law in the immiscible lattice gas [9.5]. Bubble radii R range from 4 to 64 lattice units. An estimate of surface tension is given by the slope of the best fitting line that passes through the origin. Error bars are smaller than the size of the symbols.

between a red and a blue particle. The color gradient **f** points to the right, indicating that if an interface were present, it would be oriented vertically. Maximization of expression (9.4) and conservation of mass, momentum, and color allows for only one outcome, in which each particle is directed parallel to the color gradient (or perpendicular to the interface) and towards particles of the same color. Aspects of color redistribution aside, the *dynamical* effect of such a collision rule is to increase as much as possible the difference between the component of the pressure normal to the interface and the component of the pressure tangential to the interface. This reduction of tangential pressure may be directly related to the surface tension, since it is a consequence, in real fluids, of the capillary force **f**$_c$ that acts to maintain tension on the interface.

To make the argument more clear, note that for an interface at rest, the momentum-balance equation (2.29) gives

$$\partial_\beta \Pi_{\alpha\beta} = 0, \qquad (9.16)$$

where $\Pi_{\alpha\beta}$ is again the momentum flux density tensor. Assume now that the interface is flat, and oriented parallel to the x-axis and perpendicular to the y-axis as shown in Figure 9.6. In this case, equation (9.16) gives

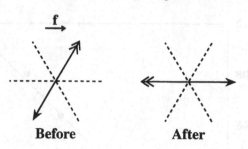

Before   **After**

Fig. 9.5. A two-body two-color zero-momentum collision in the ILG. Dashed
lines imply that no particle is present. The color gradient vector **f** points to
the right, indicating that neighboring sites to the right are populated by mostly
red particles (bold arrows) while neighboring sites to the left are populated by
mostly blue particles (double arrows). The outcome of the collision increases as
much as possible the difference between the component of pressure parallel to **f**
and the component of pressure normal to **f**.

the specific relations

$$\partial_x \Pi_{xx} + \partial_y \Pi_{xy} = 0 \qquad (9.17)$$
$$\partial_x \Pi_{yx} + \partial_y \Pi_{yy} = 0. \qquad (9.18)$$

Because the interface is parallel to the x-axis, Π_{xy} is invariant under the
reflection $x \to -x$. As discussed in Appendix A.2, such a symmetry gives
$\Pi_{xy} = \Pi_{yx} = 0$. Then from equations (9.17) and (9.18) we find

$$\partial_x \Pi_{xx} = 0 \qquad \text{and} \qquad \partial_y \Pi_{yy} = 0. \qquad (9.19)$$

Since the second of these two relations must be true arbitrarily far from
the interface, we have the solution $\Pi_{yy} = p$, where p is the isotropic
pressure. There is no explicit solution for Π_{xx}. Because it is a force
directed parallel to the interface that varies only in the perpendicular
direction, we call Π_{xx} the *tangential pressure* p_t.

   To see the importance of p_t, consider once again a curved interface,
but now allow it to have a thickness h as shown in Figure 9.7. Rather
than having the capillary force f_c pulling an infinitesimally thin interface
outward as in Figure 9.3, we now have the tangential pressure $p_t(r)$, where

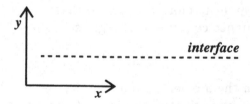

Fig. 9.6. A flat interface oriented parallel to the x-axis.

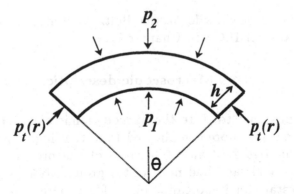

Fig. 9.7. An interface of thickness h and radius of curvature R (measured from the center of the interface). Pressures p_1 and p_2 act on the concave and convex sides of the interface, respectively. The pressure $p_t(r)$ acts on the other two sides as shown.

r is the radial coordinate, pushing a thick interface inward. Taking the interface to be centered at $r = R$, the force balance (9.13) is then modified such that

$$\int_{-\theta}^{\theta} [p_2(R + h/2) - p_1(R - h/2)] \cos\phi \, d\phi - \int_{R-h/2}^{R+h/2} 2p_t(r) \sin\theta \, dr = 0.$$
$$(9.20)$$

Carrying out the first integration and rearranging terms, we obtain

$$R(p_1 - p_2) = \int_{R-h/2}^{R+h/2} \left[\frac{p_1 + p_2}{2} - p_t(r) \right] dr. \qquad (9.21)$$

From Laplace's law (9.14) we identify the left-hand side to be the surface tension σ, and note that this remains true regardless of the size of R. Taking the limit $R \to \infty$, the interface becomes flat and we have $p_1 + p_2 = 2p$. Substituting into equation (9.21) and taking the interface to be infinitely thick, we obtain the *mechanical definition* of surface tension,

$$\sigma = \int_{-\infty}^{\infty} [p - p_t(y)] \, dy. \qquad (9.22)$$

We have taken the integration variable to be y to emphasize that this relation is valid for any flat interface perpendicular to the y direction, such as in Figure 9.6. The argument leading to equation (9.22) follows the same line in 3D and leads to the same result.

Now look back at Figure 9.5. We see, as we have already indicated, that p_t is reduced as a consequence of the ILG collision rule. If we know by how much, then the mechanical definition (9.22) allows us to calculate σ. Indeed, equation (9.22) forms the basis for the theoretical prediction of

surface tension, as we show for lattice-Boltzmann models in the following
chapter and later for ILG's in Chapter 17.

9.4   Macroscopic description

We are now prepared to state the hydrodynamic equations of the ILG.
These are somewhat more complicated than the single phase models we
studied in Chapters 2, 3 and 4 because of the presence of interfaces.
However we have already laid most of the groundwork.

First we restate an important point: for a pure blue or red phase,
the ILG's microdynamics are precisely the same as that of the random-
collision seven-bit rest-particle model of Section 3.6. Thus in the pure
phases we have the same Euler equation (3.55) we derived previously.
Moreover, the viscous term in the Navier-Stokes equation has the viscosity
ν associated with the random 7-velocity model of Table 4.1, and the
general expression for ν may be obtained from calculating the eigenvalue
λ that appears in equation (4.66). The mass-conservation or continuity
equation (3.58) is also valid in each phase.

At the interfaces between each phase we have *jump conditions*. We write
$[X] = X^{(1)} - X^{(2)}$ to signify the difference between the limit of quantity X
when the interface is approached from side 1 and the corresponding limit
when the interface is approached from side 2. We have already derived
the most important jump condition—the one related to surface tension.
From our previous developments above, the formal expression is

$$[\Pi_{\alpha\beta}]n_\beta = n_\alpha \sigma \kappa, \qquad (9.23)$$

where n_α is the α-component of the unit normal to the interface pointing
towards side 2, and the curvature $\kappa = 1/R$ is taken to be positive when
side 1 of the interface is concave. Because equation (9.23) describes a
jump in stress, it is called a *dynamic* jump condition.

There are also two *kinematic* jump conditions related to the motion of
the interface. The first states that the fluid velocities on each side of the
interface are equal:

$$[\mathbf{u}] = 0. \qquad (9.24)$$

The second condition states that the velocities in the direction normal to
the interface are equal to the normal velocity V_I of the interface itself:

$$u_\alpha^{(1)} n_\alpha = u_\alpha^{(2)} n_\alpha = V_I. \qquad (9.25)$$

The validity of equation (9.25) derives from the conservation of color
on each side of the interface. However, V_I is understood to be the average velocity of the lattice gas particles, and *not* the rescaled velocity

$g(\rho)\mathbf{u}$ defined to recover the Galilean-invariant Navier-Stokes equations (see Sections 2.7 and Appendix D). Thus ILG simulations correspond to real flows only at low Reynolds number, for which contributions from the nonlinear terms in equation (3.55) are negligible. For high Reynolds number flows, the ILG we have just described must be modified so that $g(\rho) = 1$.

The absence of a jump in velocity, equation (9.24), may be addressed through symmetry arguments. Consider a uniform flow in the x-direction parallel to a flat interface, such that there is a constant velocity $u^{(i)}$ in each phase i. Now suppose there is a discontinuity at the interface such that $u^{(1)} > u^{(2)}$, as might occur, for example, if the interface had vanishing viscosity allowing one phase to slip past the other. The symmetry of the red and blue phases indicates that if the case $u^{(1)} > u^{(2)}$ can exist, then so must another solution with the velocities $u^{(1)}$ and $u^{(2)}$ interchanged. However, the solution for flow parallel to the interface is probably unique (as indicated, for example, from numerical simulations), so it appears that we may only have the single solution $u^{(1)} = u^{(2)}$. Such an argument does not hold, however, for asymmetric phases with different viscosities, as may be constructed, for example, by using different collision rules in each phase. In this case non-physical jumps may indeed exist.

9.5 Exercises

9.1 What is the minimum number of distinct angle codes f_* that must be specified so that the choice of $\mathbf{r}'$ and $\mathbf{b}'$ made by maximizing expression (9.4) is unaffected by the approximation $\mathbf{f} \to \hat{\mathbf{f}} \approx f_*$?

9.2 Show that Laplace's law for a spherical bubble of radius R in 3D is $p_1 + p_2 = 2\sigma/R$, i.e., a slight modification of the 2D result given by equation (9.14).

9.3 Derive equation (9.22) for the case of a planar interface in a 3D fluid.

9.6 Notes

The immiscible lattice gas was introduced in:

9.1 Rothman, D. H. and Keller, J. (1988). Immiscible cellular-automaton fluids. *J. Stat. Phys.* **52**, 1119–1127.

ILG models without explicit dependence on neighboring sites have been proposed in:

9.2 Rem, P. C. and Somers, J. A. (1989). Cellular automata on a transputer network. In Monaco, R., editor, *Discrete Kinetic Theory, Lattice-Gas Dynamics, and Foundations of Hydrodynamics*, pages 268–275 (World Scientific).

9.3 Somers, J. A. and Rem, P. C. (1991). Analysis of surface tension in two-phase lattice gases. *Physica D* **47**, 39–46.

9.4 Chen, S., Doolen, G., Eggert, K., Grunau, D., and Loh, E. (1991). Local lattice-gas model for immiscible fluids. *Phys. Rev. A* **43**, 7053–7056.

These papers use colored "holes" in addition to colored particles. The paper by Somers and Rem includes a theoretical analysis of the surface tension. Interfaces in the ILG were then further investigated in:

9.5 Adler, C., d'Humières, D., and Rothman, D. H. (1994). Surface tension and interface fluctuations in immiscible lattice gases. *J. Physique I France* **4**, 29–46.

Figure 9.2 is from:

9.6 Rothman, D. H. (1992). Simple models of complex fluids. In Mareschal, M. and Holian, B., editors, *Microscopic Simulations of Complex Hydrodynamics*, pages 221-238 (Plenum Press, New York).

The mechanical definition of surface tension, equation (9.22), is derived differently in:

9.7 Rowlinson, J. and Widom, B. (1982). *Molecular Theory of Capillarity* (Clarendon Press, Oxford).

The continuum mechanics of two-phase flow concerns itself not only with the definition of jump conditions but also the derivation of constitutive equations, as may be seen in, for example:

9.8 Drew, D. A. (1983). Mathematical modeling of two-phase flow. *Ann. Rev. Fluid Mech.* **15**, 261–291.

10

Lattice-Boltzmann method
for immiscible fluids

The Boolean immiscible lattice gas of the previous chapter has a relatively simple representation as a lattice-Boltzmann method. This simplicity brings several advantages, perhaps the most important of which is the relative ease of implementation in three dimensions. But that is not all. We will now be able to write explicit equations for the collision operator, from which a prediction of the surface tension is straightforward to obtain.

This chapter may be read independently of the previous chapter on ILG's. In our discussion of surface tension here, however, we will largely omit the physical discussion we have given previously. We will also make reference to the macrodynamical behavior of ILG's, which, aside from some issues such as the Galilean invariance, applies equally well to the Boltzmann method.

10.1   Evolution equations

As with the lattice-Boltzmann method for miscible mixtures in Chapter 8, we again define red and blue populations R_i and B_i such that $R_i + B_i = N_i$, and do not include any rest particles. Furthermore, mass, momentum, and color are conserved. It is more convenient now, however, to treat the red and blue populations explicitly rather than working only with the color fraction θ.

The main idea is to split the evolution of the populations into four steps. The first step is the usual collision equation. Using the relaxation Boltzmann model of equation (6.8), we have

$$N_i'(\mathbf{x}, t) = N_i(\mathbf{x}, t) + \lambda[(N_i(\mathbf{x}, t) - N_i^{(P)}(\mathbf{x}, t)], \qquad (10.1)$$

where $N_i^{(P)}$ is given by equations (6.3) and (6.4), and the relaxation parameter λ is again related to the viscosity according to equation (6.25).

The second step is responsible for the creation of surface tension. After computing the local color gradient

$$\mathbf{f}(\mathbf{x},t) = \sum_i \mathbf{c}_i \sum_j [R_j(\mathbf{x}+\mathbf{c}_i,t) - B_j(\mathbf{x}+\mathbf{c}_i,t)] \qquad (10.2)$$

we add a perturbation to the populations N_i' such that

$$N_i''(\mathbf{x},t) = N_i'(\mathbf{x},t) + A|\mathbf{f}(\mathbf{x},t)| \left[\frac{(\mathbf{c}_i \cdot \mathbf{f})^2}{\mathbf{f}\cdot\mathbf{f}} - \frac{1}{2} \right]. \qquad (10.3)$$

Here A is a parameter chosen to set the magnitude of the surface tension and the form of the perturbation is valid for both the hexagonal and FCHC lattices. To better understand the perturbation, it is useful to specialize to the two-dimensional case on the hexagonal lattice. Equation (10.3) then takes the more transparent form

$$N_i''(\mathbf{x},t) = N_i'(\mathbf{x},t) + \frac{A}{2}|\mathbf{f}(\mathbf{x},t)| \cos 2\Delta, \qquad (10.4)$$

where Δ is the angle between $\mathbf{f}$ and $\mathbf{c}_i$, i. e.,

$$\Delta = \phi - i\pi/3, \qquad (10.5)$$

where ϕ is the angle between $\mathbf{f}$ and the x-direction ($\mathbf{c}_6$). We see then that the perturbation is chosen to remove mass from populations moving parallel to the interface and add mass to populations moving in the normal direction while conserving the total mass and momentum. As described in Section 9.3, such a redistribution of momentum acts to decrease the tangential pressure p_t and thus create surface tension, which we expect here to be proportional to the parameter A.

The third step of the algorithm redistributes the color among the populations N_i'' to minimize the diffusion of one color into the other. This amounts to choosing R_i'' and B_i'' such that

$$\sum_i (R_i'' - B_i'') \mathbf{c}_i \cdot \mathbf{f} \qquad (10.6)$$

is maximized subject to the conservation of the red mass,

$$\sum_i R_i''(\mathbf{x},t) = \sum_i R_i(\mathbf{x},t), \qquad (10.7)$$

and the conservation of the total mass on each link,

$$R_i''(\mathbf{x},t) + B_i''(\mathbf{x},t) = N_i''(\mathbf{x},t). \qquad (10.8)$$

The fourth and final step of the method is the usual propagation step:

$$R_i(\mathbf{x}+\mathbf{c}_i,t+1) = R_i''(\mathbf{x},t), \qquad (10.9)$$
$$B_i(\mathbf{x}+\mathbf{c}_i,t+1) = B_i''(\mathbf{x},t). \qquad (10.10)$$

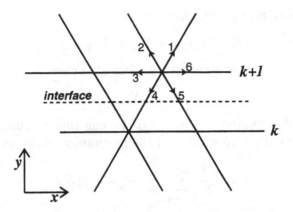

Fig. 10.1. An interface oriented parallel to the x-axis. Lattice directions c_1 through c_6 are shown a site whose discrete vertical coordinate is $k + 1$.

A conceptual difference between this method and the Boolean ILG of Chapter 9 is the explicit separation of the redistribution of mass to create surface tension and the redistribution of color to minimize diffusivity and make interfaces sharp. Another important difference is that the Boltzmann method allows the surface tension to be better understood theoretically, as we now show.

10.2 Calculation of the surface tension

For pedagogical reasons, we specialize now to the case of the hexagonal lattice. The prediction of surface tension for FCHC models follows in a similar fashion, as we point out at the end of this section. In both cases we limit our discussion to models without rest particles.

To fix ideas, consider an interface parallel to the x-axis, as shown in Figure 10.1. We assume that the system is in steady state. It is then invariant with respect to a translation in x, so it suffices to consider only the discrete vertical coordinate $k = 2y/\sqrt{3}$. From equation (9.19) we know that the component of the pressure normal to the interface, Π_{yy}, must everywhere equal the isotropic pressure p. We thus have, for every k,

$$\Pi_{yy} = \sum_i N_{ik} c_{iy} c_{iy} = p = 3f, \tag{10.11}$$

where N_{ik} is the value of N_i in the kth row and f is once again the reduced density $\rho/6$ (far from the interface), which should not be confused with the color gradient $\mathbf{f}$. Symmetry under the reflection $x \rightarrow -x$ and the steady-state condition (9.19) then gives

$$N_{1k} = N_{2k} = N_{4k} = N_{5k} = f \tag{10.12}$$

for all k. By the same symmetry, we must also have

$$N_{3k} = N_{6k}. \tag{10.13}$$

Once we have determined N_{3k} and N_{6k}, we can then calculate the surface tension from the discrete version of the mechanical definition (9.22):

$$\sigma = \frac{\sqrt{3}}{2} \sum_{k=-\infty}^{\infty} \sum_{i} (c_{iy}^2 - c_{ix}^2) N_{ik}. \tag{10.14}$$

So what are N_{3k} and N_{6k}? Remarkably, we can obtain a closed form solution for σ without explicitly calculating these populations. To see this, it is first useful to note that the surface tension step (10.3) (or, more specifically, equations (10.4) and (10.5), where $\phi = \pi/2$) may be written in the simpler form

$$N_{ik}'' = N_{ik}' - A|\mathbf{f}_k| Q_{ixx}, \tag{10.15}$$

where Q_{ixx} is given by equation (3.43). We expect a steady solution given by equation (10.12) for the four known populations and

$$N_{3k} = N_{6k} = f - \epsilon_k \tag{10.16}$$

for the two unknown populations, where ϵ_k is an unknown deviation from isotropy. Equation (10.14) then simplifies to a sum over the ϵ_k's:

$$\sigma = \sqrt{3} \sum_{k=-\infty}^{\infty} \epsilon_k. \tag{10.17}$$

To make further progress, we note that we may combine equations (10.12) and (10.16) into the single steady solution

$$N_{ik} = f - \frac{1}{3}\epsilon_k - \frac{4}{3}\epsilon_k Q_{ixx}. \tag{10.18}$$

Noting that the first two terms on the right-hand side belong to the pseudo-equilibrium distribution $N_i^{(P)}$, we see that the collision step, equation (10.1), takes the form

$$N_{ik}' = N_{ik} - \frac{4}{3}\lambda \epsilon_k Q_{ixx}. \tag{10.19}$$

Then from equation (10.15) we obtain

$$N_{ik}'' = N_{ik} - \left(\frac{4}{3}\lambda \epsilon_k + A|\mathbf{f}_k|\right) Q_{ixx}. \tag{10.20}$$

From equations (10.12) and (10.13), the steady state is simply $N_{ik}'' = N_{ik}$, which gives

$$-\frac{4}{3}\lambda\epsilon_k = A|\mathbf{f}_k|. \tag{10.21}$$

Equation (10.17) may now be written as a sum over the magnitudes of the color gradients:

$$\sigma = \frac{-3A\sqrt{3}}{4\lambda} \sum_{k=-\infty}^{\infty} |\mathbf{f}_k|. \tag{10.22}$$

How can we evaluate such a sum? The solution could not be simpler. Setting $\rho_k = \sum_i N_{ik}$ and $\theta_k = \sum_i R_{ik}/\rho_k$, we may calculate $|\mathbf{f}_k|$ from equation (10.2):

$$|\mathbf{f}_k| = \sqrt{3}[(2\theta_{k+1} - 1)\rho_{k+1} - (2\theta_{k-1} - 1)\rho_{k-1}]. \tag{10.23}$$

Using equations (10.12) and (10.16) and summing over k, we obtain

$$\sum_{k=-m}^{k=m} |\mathbf{f}_k| = 2\sqrt{3} \sum_{k=-m}^{k=m} [6f(\theta_{k+1} - \theta_{k-1}) + (\epsilon_{k+1} - \epsilon_{k-1}) -$$
$$2(\theta_{k+1}\epsilon_{k+1} - \theta_{k-1}\epsilon_{k-1})]. \tag{10.24}$$

Here we have temporarily replaced the infinite limits on the summation index k with the finite limits $-m$ and m. The sum then simplifies, since each difference contained between parentheses contributes to the sum only at the end points, yielding

$$\sum_{k=-m}^{k=m} |\mathbf{f}_k| = 2\sqrt{3}[6f(\theta_{m+1} + \theta_m - \theta_{-m-1} - \theta_{-m}) +$$
$$(\epsilon_{m+1} + \epsilon_m - \epsilon_{-m-1} - \epsilon_{-m}) -$$
$$2(\theta_{m+1}\epsilon_{m+1} + \theta_m\epsilon_m - \theta_{-m-1}\epsilon_{-m-1} - \theta_{-m}\epsilon_{-m})]. \tag{10.25}$$

Some simple physical considerations now allow us to complete the analysis. First, we expect that the perturbation $\epsilon_k \to 0$ as $k \to \pm\infty$, i. e., far from the interface. Second, in the absence of any diffusive mixing (a consequence of the recoloring step), we expect the fluids to be pure red or pure blue far from the interface. Taking the fluid to be red for large k, this means $\theta_k \to 1$ as $k \to \infty$ and $\theta_k \to 0$ as $k \to -\infty$. Thus as $m \to \infty$, the sum (10.25) may be trivially evaluated, and the surface tension (10.22) becomes

$$\sigma = \frac{-54Af}{\lambda}. \tag{10.26}$$

The analogous calculation for the FCHC lattice follows the same argument. For a two-dimensional interface parallel to the z-axis, the symmetries of the FCHC lattice and the perturbation (10.3) give the steady solution

$$N_{ik} = f - \frac{1}{2}\epsilon_k + \epsilon_k Q_{izz} \qquad (10.27)$$

instead of equation (10.18). Construction and evaluation of the summation as done above then yields the result

$$\sigma = \frac{-4608Af}{\lambda}. \qquad (10.28)$$

10.3   Theory versus simulation

Rather than demonstrating Laplace's law for a bubble as we did in Chapter 9, in Figure 10.2 we show an example of a measurement of the surface tension of a flat interface. The interface was simulated with a lattice-Boltzmann method on the FCHC lattice, and surface tension was calculated according to the mechanical definition, equation (9.22). The measurements are compared to the theoretical prediction (10.28), and one sees that the match between theory and simulation is excellent. A similar fit to the theoretical prediction is obtained from simulations of bubbles.

This is not quite the end of the story, however. Figure 10.3 shows a picture of the velocity field $u(x)$ at a particular time step in and around a typical two-dimensional bubble simulated with the lattice-Boltzmann method on a hexagonal lattice. In the real world one would expect the velocity to be zero everywhere. What we find here, however, are *spurious currents* that peak in amplitude near the surface of the bubble and decay approximately exponentially with distance from the interface. As one can see from Figure 10.3, there is a six-fold symmetry in the pattern, revealing immediately that the spurious currents are related to the discrete nature of the space in which the bubble resides—a hexagonal lattice.

Empirical studies show that the magnitude v_{max} of the most egregious spurious currents increases linearly with surface tension and decreases linearly with the inverse of viscosity. This suggests the scaling

$$v_{max} \propto C^{-1}, \qquad (10.29)$$

where $C = \rho \nu U / \sigma$ is the dimensionless *capillary number*, with U a typical velocity scale for the flow. Fortunately, the constant of proportionality is only of order 10^{-2}, so the spurious currents are often negligible in simulations of interfacial flows. But care must be taken, in the form of appropriate validations, to make sure that these currents do not cause erroneous conclusions to be made from simulations.

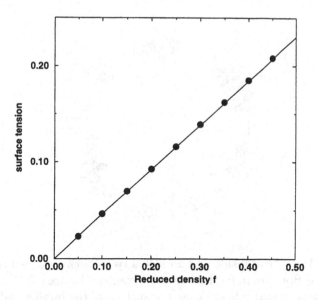

Fig. 10.2. Comparison of the theoretical prediction for surface tension given by equation (10.28) with results from simulations of the lattice-Boltzmann method for immiscible fluids on the FCHC lattice. Measurements of surface tension were made using a flat interface. In the example shown, the parameter $A = 10^{-4}$ and $\lambda = -1$.

We close this brief section by noting that macroscopic behavior of the lattice-Boltzmann method for immiscible fluids is expected to be nearly identical to its Boolean counterpart, the ILG. Thus the discussion of jump conditions in Chapter 9.4 applies equally well here. Viscous hydrodynamics is also once again expected to be identical to that of the corresponding single-phase model. For the case we have just discussed, this means that the single-phase fluid dynamics of the relaxation-Boltzmann model (6.8) will govern the behavior of the immiscible-fluid model away from interfaces. Recall, however, that we showed in Section 6.2 that the Boltzmann method may be easily engineered to yield a Galilean-invariant Euler equation. This is a significant advantage compared to the Boolean ILG if one wishes to study interfacial flows with non-negligible Reynolds number.

10.4 Exercises

10.1 Carry out the calculation of the FCHC surface tension given by equation (10.28).

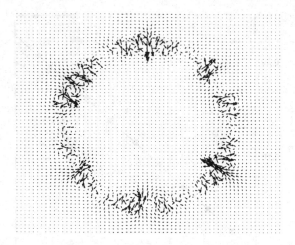

Fig. 10.3.　Velocity field inside and outside a two-dimensional bubble simulated with the lattice-Boltzmann method on the hexagonal lattice [10.3]. The currents with the greatest magnitude are near the surface of the bubble and show a six-fold symmetry.

10.5　Notes

The lattice-Boltzmann method for immiscible fluids was introduced in:

10.1 Gunstensen, A. K., Rothman, D. H., Zaleski, S., and Zanetti, G. (1991). A lattice-Boltzmann model of immiscible fluids. *Phys. Rev. A.* **43**, 4320–4327.

This paper also shows that interfaces simulated with the Boltzmann method correctly satisfy the capillary-wave dispersion relation. The method was later extended to 3D in:

10.2 Gunstensen, A. K. and Rothman, D. H. (1992). Microscopic modeling of immiscible fluids in three dimensions by a lattice-Boltzmann method. *Europhys. Lett.* **18**, 157–161.

Both of these papers form part of:

10.3 Gunstensen, A. K. (1992). *Lattice-Boltzmann Studies of Multiphase Flow Through Porous Media.* Ph. D. thesis, Massachusetts Institute of Technology.

Gunstensen's prediction of the surface tension coefficient followed a different argument than that given in this chapter. The calculation presented here is due to D. d'Humières and is reported in:

10.4 Giraud, L., d'Humières, D., Paul, B., and Rothman, D. H. (1996). Exact solutions for the immiscible lattice-Boltzmann models with plane interfaces (preprint).

Several variations of Gunstensen's method have been proposed. These include:

10.5 Grunau, D., Chen, S., and Eggert, K. (1993). A lattice-Boltzmann model for multiphase fluid flows. *Phys. Fluids A* **5**, 2557–2562.

10.6 Shan, X. and Chen, H. (1994). Lattice-Boltzmann model for simulating flows with multiple phases and components. *Phys. Rev. E* 47, 1815–1819.

10.7 Orlandini, E., Swift, M. R., Yeomans, J. M. (1995). A lattice-Boltzmann model of binary-fluid mixtures. *Europhys. Lett.* **32**, 463–468.

The paper by Orlandini et al. is notable for its inclusion of the free-energy functional (16.2) in order to obtain behavior compatible with a thermodynamic description of a fluid mixture. They also report a substantial reduction in the magnitude of the spurious currents shown in Figure 10.3. Their work derives from an earlier paper that uses a similar technique to construct a lattice-Boltzmann method for a liquid-gas mixture:

10.8 Swift, M. R., Osborn, W. R., and Yeomans, J. M. (1995). Lattice-Boltzmann simulation of nonideal fluids. *Phys. Rev. Lett.* **75**, 830–833.

11
Immiscible lattice gases
in three dimensions

A three-dimensional Boolean immiscible lattice gas presents a formidable challenge. Not only do we have the original complication of 2^{24} possible single-phase states at each site of the FCHC lattice, but with two colors the configuration space becomes much larger: $3^{24} \approx 10^{11}$ possible two-color states!

However, as we have already seen in Chapter 5, some modestly clever geometric insight goes a long way, and one need not be a professional geometer to play this game. Indeed, the model we discuss below is based on a trick so simple that we call it the *dumbbell method*. As we shall see, our new dumbbells have much in common with the earlier dumbbells of Chapter 5, the only real change being a bit of color.

11.1   Four familiar steps

The microdynamics of the 3D ILG bears a strong resemblance to the analogous Boltzmann model we described in Chapter 10, being composed once again of four distinct steps. Step one is again the collision step. From the colored states r_i and b_i we form

$$n_i = r_i + b_i \in \{0, 1\}, \tag{11.1}$$

and perform the usual Boolean collision such that

$$n_i'(\mathbf{x}, t) = n_i(\mathbf{x}, t) + \Delta_i[\mathbf{n}(\mathbf{x}, t)]. \tag{11.2}$$

Here we are now on the FCHC lattice, so the velocity index i runs from 1 to 24. The collision operator Δ_i may be any of the simple operators discussed in Chapter 5, or one of the more elaborate, optimized operators discussed in the notes of that chapter. As in the Boltzmann model, the choice of the single-phase collision operator enters the dynamics only through transport coefficients such as the viscosity, or material coefficients such as the surface tension (via the eigenvalue λ of the linearized operator).

Step two is the surface tension step. As usual, we need the color gradient

$$\mathbf{f}(\mathbf{x}, t) = \sum_{i=1}^{24} \mathbf{c}_i \sum_{j=1}^{24} [r_j(\mathbf{x} + \mathbf{c}_i, t) - b_j(\mathbf{x} + \mathbf{c}_i, t)]. \tag{11.3}$$

We then change the post-collision state n_i' in a manner analogous to the Boltzmann surface tension step of equation (10.3). Working only with the uncolored states n_i and the unit color gradient $\hat{\mathbf{f}} = \mathbf{f}/|\mathbf{f}|$, we create a new state n_i'' such that

$$\sum_{i=1}^{24} n_i'' |\hat{\mathbf{f}} \cdot \mathbf{c}_i| \tag{11.4}$$

is maximized, subject to the constraints of mass conservation,

$$\sum_{i=1}^{24} n_i'' = \sum_{i=1}^{24} n_i', \tag{11.5}$$

and momentum conservation,

$$\sum_{i=1}^{24} n_i'' \mathbf{c}_i = \sum_{i=1}^{24} n_i' \mathbf{c}_i. \tag{11.6}$$

Maximization of the expression (11.4) enhances the pressure normal to an interface at the expense of pressure tangential to the interface, and thus results in surface tension on the interface. To make such a scheme tractable, we need to find a clever way to perform the maximization so that it can be reduced to no more than a few table-lookup steps in a computer algorithm. Indeed, we will only *approximately* maximize (11.4). This approximation, which more than suffices to create significant surface tension, is based on some tricks which we describe in more detail below. First, however, we complete the description of the microdynamics.

Step three is the recoloring step. Once the new state n_i'' has been constructed, colors r_i'' and b_i'' are reassigned to particles to minimize the diffusion of each species into the other. Specifically, r_i'' and b_i'' are chosen to maximize

$$\mathbf{q}(\mathbf{r}'', \mathbf{b}'') \cdot \hat{\mathbf{f}}(\mathbf{x}, t), \tag{11.7}$$

where

$$\mathbf{q}(\mathbf{r}'', \mathbf{b}'') = \sum_{i=1}^{24} \mathbf{c}_i (r_i'' - b_i''), \tag{11.8}$$

is the definition of the color flux on the FCHC lattice, and the number of red particles is conserved:

$$\sum_{i=1}^{24} r_i'' = \sum_{i=1}^{24} r_i. \tag{11.9}$$

From equation (11.1) we see that conservation of red particles implies conservation of blue particles because the state n_i'' is unchanged in the recoloring step.

The final step in the microdynamics of the 3D Boolean ILG is simple propagation:

$$r_i(\mathbf{x} + \mathbf{c}_i, t + 1) = r_i''(\mathbf{x}, t), \tag{11.10}$$
$$b_i(\mathbf{x} + \mathbf{c}_i, t + 1) = b_i''(\mathbf{x}, t). \tag{11.11}$$

We have thus outlined the basic algorithm. Steps one and four, single-phase collision and particle propagation, are performed just as they are in single-phase models and require no further comment. Steps two and three, however, require some new ideas so that they may be implemented with lookup tables that are not impractically large. These ideas are based on the construction of the dumbbells of Chapter 5 and a tiling of the sphere that allows an efficient approximation of $\hat{\mathbf{f}}$ on the FCHC lattice. We discuss each in turn below.

11.2   The dumbbell method

Suppose that we know $\hat{\mathbf{f}}$ with whatever accuracy we wish. How, then, should we choose the new states n_i'' so that expression (11.4) is maximized?

A precise maximization of (11.4) would require construction of a lookup table based on the 2^{24} possible input states and would thus typically be impractical. But really what we desire is just an approximate maximization, or, more precisely stated in the language of Chapters 9 and 10, a systematic perturbation of the collisions that makes the pressure locally anisotropic near an interface. This is easily achieved by a modification of the 12-bit dumbbell collisions described in Chapter 5.3.

For completeness we recall some definitions given in our previous discussion. We index the 24 directions so that $\mathbf{c}_{i+12} = -\mathbf{c}_i$. From the post-collision state $\mathbf{n}'$ we then construct two 12-bit states corresponding to the directions with low and high indices, respectively:

$$\begin{aligned} l_i(\mathbf{n}') &= n_i' \\ h_i(\mathbf{n}') &= n_{i+12}' \end{aligned} \quad \text{for } i = 1, \ldots, 12. \tag{11.12}$$

The set of *dumbbells* $\mathbf{d}(\mathbf{n}') = (d_i)_{i \in [1,12]}$ is then all those pairs of opposite directions that are both occupied. Taking d_i to be one if a dumbbell exists and zero if it does not, the appropriate expression is

$$d_i(\mathbf{n}') = h_i\, l_i. \tag{11.13}$$

Of equal interest are the empty pairs where $l_i = h_i = 0$.

Whereas in Chapter 5 we constructed an isotropic collision scheme in which dumbbells and empty pairs were randomly exchanged, our interest here is to define such an exchange that creates surface tension. Thus we now exchange dumbbells and empty pairs so that the new set of dumbbells $(d_i)_i \to (\tilde{d}_i)_i$ maximizes

$$\sum_{i=1}^{12} \tilde{d}_i |\hat{\mathbf{f}} \cdot \mathbf{c}_i|. \tag{11.14}$$

As in our earlier construction, such an exchange automatically satisfies the restriction that mass and momentum be conserved. The state $\mathbf{n}''$ of the lattice after this perturbation has been performed is then given by

$$n_i'' = \begin{cases} \tilde{l}_i & \text{for } i \le 12 \\ \tilde{h}_{i-12} & \text{for } i > 12 \end{cases}. \tag{11.15}$$

We now turn to the specification of the unit gradient vector $\hat{\mathbf{f}}$.

11.3 Symmetric tiling

As in the case of the two-dimensional ILG of Chapter 9, it is convenient to work with a discrete approximation of $\hat{\mathbf{f}}$. Whereas in two dimensions this just amounts to uniform angular increments around the unit circle, in three dimensions the problem is considerably more subtle. We instead require a tiling of the unit sphere. Moreover, such a tiling should have the cubic symmetry of the three-dimensional projection of the FCHC lattice. Fortunately an elegant solution exists.

The use of lattice symmetries proceeds by first breaking up the unit sphere into triangular tiles, each of the same size and shape but possibly of different orientation. These tiles, shown in Figure 11.1, are constructed by drawing on the sphere the three great circles C_x, C_y, and C_z in the planes normal to the x, y, and z axes, respectively. The six great circles which bisect the angles between each pair of C_x, C_y, and C_z are also drawn. This divides the sphere into 48 tiles with the required symmetries.

This tiling allows us to define a discrete version of $\hat{\mathbf{f}}$. First, we determine the tile to which $\hat{\mathbf{f}}$ points. (If $\hat{\mathbf{f}}$ points to an edge or corner of a tile, we need only determine one of the tiles with that edge or corner.) Then one of four discrete gradient codes f_* is chosen. The four f_* correspond

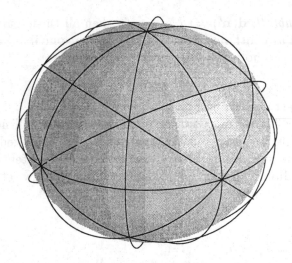

Fig. 11.1. The unit sphere is tiled with 48 identical triangles as shown, and the unit color gradients $\hat{\mathbf{f}}$ are discretized to either a corner or the center of a triangle [11.2]. The corners of the tiles are permutations of $(\pm 1, 0, 0)$, where eight tiles meet; permutations of $(\pm 1, \pm 1, 0)/\sqrt{2}$, where four tiles meet; and permutations of $(\pm 1, \pm 1, \pm 1)/\sqrt{3}$, where six tiles meet.

to the center of the tile and its three corners; the particular f_* chosen corresponds to the one that is closest to $\hat{\mathbf{f}}$. All told, there are 74 possible values for f_* on the unit sphere.

The beauty of the method we have just described is that it allows us to reduce a large number of possible color gradients to just four "archetypal" unit gradients. Why are there just four? Because each tile may be transformed to the position of one of the others by application of a lattice symmetry operation $\mathbf{R}$. Thus we need only consider one of the tiles, which we designate the *archetype*. Then once we find the tile associated with $\hat{\mathbf{f}}$, the appropriate lattice symmetry operation $\mathbf{R}$ is used to move $\hat{\mathbf{f}}$ into the archetypal tile, and simultaneously to make the transformation $\mathbf{n}' \to \mathbf{R}\mathbf{n}'$. The nearest of the four archetypal gradients is then used to perform the table-lookup operations necessitated by the surface-tension and recoloring steps, yielding $\mathbf{R}\mathbf{n}''$ and $\mathbf{R}\mathbf{r}''$. Finally, the inverse transformation $\mathbf{R}^{-1}$ is applied to these sets, returning them to the original orientation and yielding $\mathbf{n}''$ and $\mathbf{r}''$.

11.4   Surface tension, phase separation, and isotropy

Having described the construction of a 3D immiscible lattice gas, we now turn to a discussion of some of its properties.

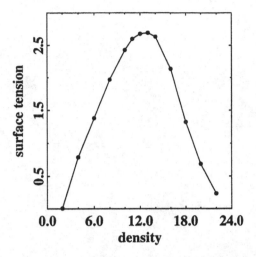

Fig. 11.2. Surface tension as a function of particle density in the 3D immiscible lattice gas [11.2]. Surface tension was measured by simulating 3D bubbles, computing the pressure difference between the inside and outside of the bubbles, and inferring the surface tension coefficient from equation (11.16).

As in the 2D ILG, the most important property is surface tension. Measurements of the pressure inside and outside bubbles as described in Chapter 9 once again show adherence to Laplace's law, which for a 3D spherical bubble reads as

$$p_1 - p_2 = 2\sigma/R. \tag{11.16}$$

Having shown one such graph already, we shall resist the temptation to show a second! Ultimately of more interest in Boolean ILG's, however, is the behavior of surface tension as a function of the particle density ρ. The results of a series of surface tension measurements with the 3D ILG at different values of ρ are shown in Figure 11.2. The form of the curve is common to Boolean ILG's in both two and three dimensions: the surface tension is zero up to a finite value of the density, it rises to a peak, and then falls to zero when the lattice is fully occupied. We will study such curves more carefully in Chapter 17 when we discuss theoretical predictions of surface tension in Boolean interface models. We point out here, however, that the most interesting physical aspect of the curve is that it indicates that the model undergoes a *phase transition* at $\rho \approx 2$, that is, a transition from zero to finite surface tension. Thus for $\rho < 2$ red-blue mixtures are stable, whereas for $\rho > 2$ they are unstable.

We see, then, that if the density is sufficiently large, 3D ILG mixtures will exhibit spontaneous phase separation just as we described for 2D ILG's in Chapter 9.2. 3D phase separation is exciting to watch (either on

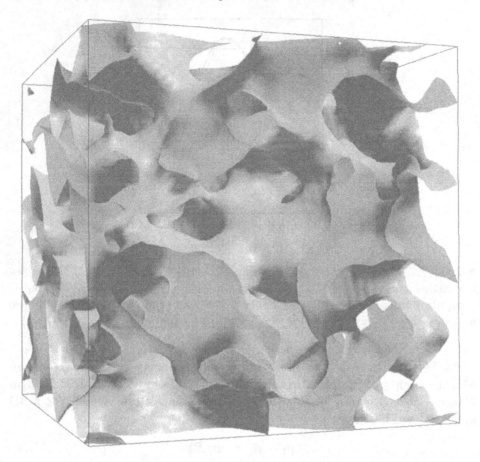

Fig. 11.3.   Phase separation in the 3D ILG [11.9]. The volume fraction of each phase is 50% and the lattice size is 128^3. Only the interfaces are shown. The snapshot shown is taken about 3000 time steps after initialization as a random mixture. The average particle density $\rho = 12$ and the boundaries are periodic in each dimension.

a computer screen or in real life!) and we cannot pass up the opportunity to show what it looks like. Thus in Figure 11.3 we see a snapshot of phase separation in the 3D ILG after the system was initialized with a mixture that contained equal amounts of red and blue particles, and in Figure 11.4 we see an analogous picture but with only 5% red particles rather than 50%.   The two pictures differ qualitatively because the way each phase fills space when the system is out of equilibrium depends strongly on the volume fraction each phase occupies. This is far from the full story however, and we shall return to a discussion of phase separation in Chapter 16.

We close this brief section on phenomenological aspects of 3D ILG's

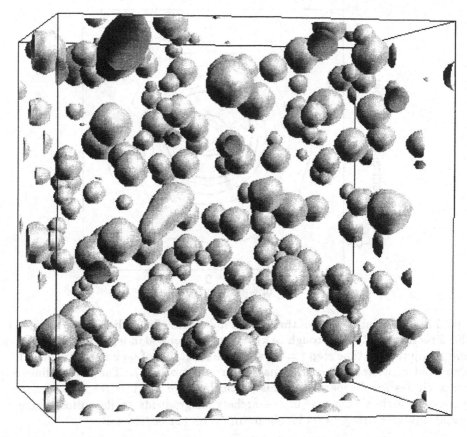

Fig. 11.4. Phase separation in the 3D ILG [11.9]. The volume fraction of the red phase is now only 5%, which results in bubbles rather than the interconnected tangle of interfaces evident in Figure 11.3. Aside from the fact that this snapshot was taken after about 6000 time steps, all other aspects of the simulation are identical to those of Figure 11.3.

with an empirical measure of the degree to which the tiling scheme produces isotropic surface tension. Indeed, as we discuss later in Chapter 17, surface tension is expected to be at least weakly anisotropic in ILG's as a result of the inherent anisotropy of the lattice. One way to evaluate the anisotropy is to simply measure its effect on the shape of a bubble, or, more generally, on the geometry of phase separation. Such a measure is made by computing the three-dimensional power spectrum of the color density field, and then averaging it over many independent simulations. Specifically, we compute

$$S(\mathbf{k}, t) = \left\langle \left| \sum_{\mathbf{x}} \psi(\mathbf{x}, t) e^{-i\mathbf{k}\cdot\mathbf{x}} \right|^2 \right\rangle \tag{11.17}$$

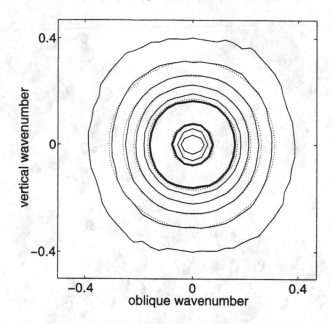

Fig. 11.5. A contour plot of the natural logarithm of an oblique slice (normal to the direction $(1, 1, 0)$) through $S(\mathbf{k})$, averaged over 600 independent simulations and computed at time step $t = 10$ after the initiation of phase separation [11.2]. The bold contours enclose the maximum values of $S(\mathbf{k})$. The dotted lines are exact circles for comparison. This particular slice has been chosen because it passes through all but one of the archetypal gradients so it is most likely to reveal any anisotropy. Axis labels are in cycles per lattice unit.

where

$$\psi(\mathbf{x}, t) = \sum_i [r_i(\mathbf{x}, t) - b_i(\mathbf{x}, t)] \qquad (11.18)$$

is the color density field and the average is taken over many independent simulations. If $S(\mathbf{k})$ is spherically symmetric, then we may conclude that the shape of the phase separating patterns is independent of orientation and that any anisotropy in surface tension is not significant enough to deform the patterns. In Figure 11.5 we show contours of a two-dimensional slice through $S(\mathbf{k})$. Because the contours are nearly circular, we conclude that the phase separating domains are approximately isotropically distributed in space. Thus the symmetric tiling scheme has not introduced any significant anisotropy.

11.5 Drag force on a bubble

In Chapter 9.4 we discussed the macroscopic behavior of immiscible lattice gases. The principal conclusion was that for slow flow and matched viscosities, one expects to recover the correct viscous hydrodynamics in addition to the classical jump conditions that result from surface tension at interfaces.

Such a prediction must be tested. One of the more interesting of the various possible validations is to ask whether the viscous drag force on a slowly moving bubble conforms to theoretical predictions. The problem is analogous to that of a spherical solid object falling slowly through a viscous fluid, but is more complex because in the case of a fluid drop the relevant fluid motions occur not only outside drop but inside as well.

Here we simply state the relevant results. When a spherical drop of radius a falls through another fluid of infinite extent, the terminal velocity of the drop—i.e., the speed at which the drag force balances the buoyancy force—is given by [11.7]

$$V(a) = \frac{4}{15} \frac{g \Delta \rho}{\mu} a^2. \tag{11.19}$$

Here $\Delta \rho$ is the density difference between the two fluids, g is the gravitational acceleration, and μ is the viscosity of both the fluid drop and the surrounding fluid.

To test the lattice gas against this theory, we would need to simulate a falling drop in a container so large that the walls of the container would have a negligible influence on the frictional drag acting against the bubble. Such a container, however, would be prohibitively large. Instead, we simulate a single falling drop in a box with periodic boundary conditions in each dimension. This is equivalent to a physical experiment in which an infinite periodic array of bubbles, each situated at the node of an imaginary cubic lattice, falls through another fluid. Theory predicts that the terminal velocity U of this array is given by a multiplicative correction to $V(a)$ that depends on the volume fraction ϕ occupied by the bubbles:

$$U(\phi; a) = \gamma(\phi) V(a). \tag{11.20}$$

Lattice-gas simulations of falling bubbles are thus tested against equation (11.20), for which $\gamma(\phi)$ is given in Ref. [11.8].

To simulate a falling drop, we place a bubble of radius a in a periodic box of linear dimension L, giving a volume fraction $\phi = 4\pi a^3/(3L^3)$. We apply a body force to both fluids in a way that maintains a net momentum of zero in the entire box. In other words, we force the bubble downward with force density f_{bub} while forcing the external fluid upward with force density f_{ext} such that $(1 - \phi)f_{ext} = -\phi f_{bub}$. We then mea-

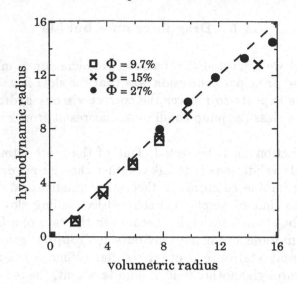

Fig. 11.6. Comparison of the measured hydrodynamic radius a_* with the specified volumetric radius a in simulations of a slowly moving cubic array of bubbles [11.5]. Three sets of data are shown, for volume fractions $\phi = 9.7\%$ (squares), $\phi = 15\%$ (×'s), and $\phi = 27\%$ (circles). Measurements match theory if they fall on the dashed line.

sure the terminal velocity U_* of the array. The measurement U_* should correspond to U in equation (11.20). However, to better understand the source of any possible errors in the jump conditions, it is more useful to compare the specified, or "volumetric," radius a against the effective, or "hydrodynamic," radius a_* that is predicted by substituting U_* for $U(\phi; a)$ in equations (11.19) and (11.20) and then solving for a_*. Note that in the computation of a_*, μ and ϕ are fixed by the simulation setup, and $g\Delta\rho = f_{bub}/(1 - \phi)$.

Figure 11.6 compares the hydrodynamic radius a_* against the volumetric radius a for three different volume fractions ϕ. One sees that for each volume fraction, $a \approx a_*$ for bubble radii ranging from 3 to 15 lattice units. Viewed in detail, the results indicate that the effective hydrodynamic radius may be systematically somewhat smaller than the volumetric radius. Such a small systematic deviation would not be surprising, as the problem of locating the effective boundary of a fluid-fluid interface is theoretically analogous to the location of solid walls discussed in Section 7.4.

11.6   Exercises

11.1 Show that the tiling scheme just described in Section 11.3 results in 74 discrete orientations of the unit gradient vector. What other alternative tiling schemes can you propose?

11.2 If you were to construct a collision table to perform the maximization of expression (11.4), how many entries would it have?

11.3 Why may the surface tension be anisotropic when the viscous stress tensor, for example, is not?

11.7 Notes

The first 3D immiscible lattice gas was reported in:

11.1 Rem, P. C. and Somers, J. A. (1989). Cellular automata on a transputer network. In Monaco, R., editor, *Discrete Kinetic Theory, Lattice-Gas Dynamics, and Foundations of Hydrodynamics*, pages 268–275 (World Scientific).

This work used colored holes in addition to colored particles.
The model described in this chapter was described in detail in:

11.2 Olson, J. F. and Rothman, D. H. (1995). A three-dimensional immiscible lattice gas: application to sheared phase separation. *J. Stat. Phys.* **81**, 199–222.

The symmetric tiling of Figure 11.1 was proposed by Olson, and is one of several technical innovations found in:

11.3 Olson, J. F.(1995). Two-fluid flow in sedimentary rock: complexity, transport, and simulation. Ph. D. thesis, Massachusetts Institute of Technology.

The same tiling may also be found in Figure 4.5a of:

11.4 Coxeter, H. S. M. (1977). *Regular Polytopes* (Dover, New York).

The calculation of bubble drag is described in:

11.5 Olson, J. F. and Rothman, D. H. (1996). Two-fluid flow in sedimentary rock: simulation, transport, and complexity. *J. Fluid Mech.*, in press.

It is also in Olson's thesis of the same title [11.3]. An analogous calculation of drag for a 3D array of hard spheres, computed by the lattice-Boltzmann method, is in:

11.6 Ladd, A. J. C. (1994). Numerical simulations of particulate suspensions via a discretized Boltzmann equation. Part 2. Numerical results. *J. Fluid Mech.* **271**, 311–340.

The drag of a single bubble is discussed in:

11.7 Lamb, H. (1932). *Hydrodynamics* (Cambridge University Press, New York).

The theoretical result for a cubic array of bubbles is due to:

11.8 Sangani, S. (1987). Sedimentation in ordered emulsions of drops at low Reynolds numbers. *ZAMP* **38**, 542–556.

Figures 11.3 and 11.4 are from:

11.9 Appert, C., Olson, J. F., Rothman, D. H., and Zaleski, S. (1995). Phase separation in a three-dimensional two-phase hydrodynamic lattice gas. *J. Stat. Phys.* **81**, 181–197.

These figures should be thought of as a pictorial introduction to our discussion of phase separation in Chapter 16.

12

Liquid-gas models

In this chapter we introduce a new kind of lattice-gas mixture. Unlike our previous mixture models, this new mixture has only one species of particle. For certain choices of parameters, however, two thermodynamic phases can coexist. One phase, which we call liquid, has a high density of particles, while the other phase, which we call gas, has a relatively low density. Interfaces form between phases spontaneously, and the model exhibits phase separation in a way similar to that of immiscible lattice gases.

Remarkably, the rich phenomenology of these *liquid-gas* models arises from a microdynamics that is surprisingly simple, both in two and three dimensions. Moreover, many of the salient theoretical results are easily derived. We begin, as usual, with a description of the microdynamics.

12.1 Interactions at a distance

Figure 12.1 illustrates the microdynamics of the two-dimensional liquid-gas model. Compared to the FHP model described in Figure 1.1, the initial state and the hopping step are precisely the same as before. There are now two collision steps, however. The first, shown in Figure 12.1c, is the same as in the FHP model. As usual, we write the outcome of this "classical" collision as

$$n_i' = n_i + \Delta_i(\mathbf{n}). \tag{12.1}$$

The second collision step, Figure 12.1d, consists of interactions at a specified range r, where here we choose $r = 2$. Lattice sites at locations $\mathbf{x}$ and $\mathbf{x} + r\mathbf{c}_i$ trade particles moving in directions $-\mathbf{c}_i$ and $\mathbf{c}_i$, respectively, if and only if both particles exist prior to the exchange *and* the exchange can be performed without creating more than one particle moving with any velocity. Figure 12.2 illustrates the rule in detail.

141

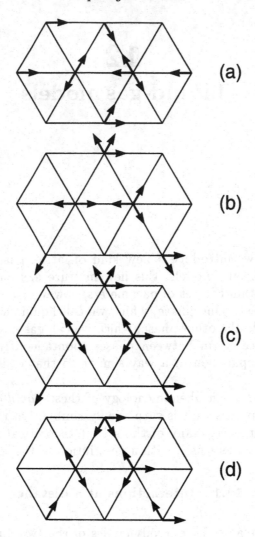

Fig. 12.1. Microdynamics of the liquid-gas model, in which the initial condition (a), the propagation step (b), and collision step (c) are precisely the same as that showed in Figure 1.1. The new, interaction, step is shown in (d). For simplicity, in this example the interaction distance $r = 2$.

To obtain an explicit expression for this "interaction at a distance," we define the Boolean variable

$$\gamma_i = \overline{n}'_i(\mathbf{x})n'_{i+3}(\mathbf{x})n'_i(\mathbf{x} + r\mathbf{c}_i)\overline{n}'_{i+3}(\mathbf{x} + r\mathbf{c}_i), \qquad (12.2)$$

where, as in Section 2.6, overbars once again are used such that $\overline{x} = 1 - x$ and the circular shift $i+3$ is defined as in equations (2.35) and (2.37). The microdynamical equation describing the complete sequence of propagation

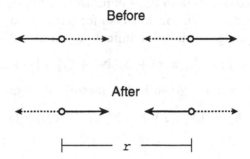

Fig. 12.2. An explicit example of an interacting collision that may occur in the liquid-gas model. Solid arrows represent particles; broken arrows represent the absence of a particle. The sites on which the interaction occurs are situated r lattice units apart.

followed by classical and interacting collisions is then

$$n_i(\mathbf{x} + \mathbf{c}_i, t + 1) = n_i'(\mathbf{x}, t) + \gamma_i - \gamma_{i+3}. \tag{12.3}$$

A little thought reveals a nice property of these interacting collisions: they may be performed in any sequence whatsoever, or, indeed, in parallel, just like the usual lattice gas collisions.

The microdynamics 12.3 may be generalized to any dimension effort-lessly. We let the indices i and $-i$ denote opposite pairs of velocities, such that $\mathbf{c}_i = -\mathbf{c}_{-i}$. The interaction term γ_i is now given by

$$\gamma_i = \overline{n}_i'(\mathbf{x}) n_{-i}'(\mathbf{x}) n_i'(\mathbf{x} + r\mathbf{c}_i) \overline{n}_{-i}'(\mathbf{x} + r\mathbf{c}_i). \tag{12.4}$$

The microdynamical equation of the liquid-gas model in any dimension is then

$$n_i(\mathbf{x} + \mathbf{c}_i, t + 1) = n_i'(\mathbf{x}, t) + \gamma_i - \gamma_{-i}. \tag{12.5}$$

That's it!

12.2 Equation of state

We begin our analysis of the liquid-gas model by calculating its equation of state for a zero-velocity equilibrium. Recall that the inviscid form of the momentum flux density tensor for a non-interacting model is given by

$$\Pi_{\alpha\beta}^{(0)} = \sum_i N_i^{eq} c_{i\alpha} c_{i\beta}, \tag{12.6}$$

where N_i^{eq} is again the Fermi-Dirac equilibrium (3.42). Interactions result in a modification of this expression. To see how, we must first make the molecular chaos, or Boltzmann approximation (Chapter 3.1), not only for particles incoming to a collision but also for particles on interacting sites. Averaging the liquid-gas microdynamical equation (12.5) then yields

$$N_i(\mathbf{x} + \mathbf{c}_i, t + 1) = N_i(\mathbf{x}, t) + \Delta_i(\mathbf{N}) + \Gamma_i[\mathbf{N'}] - \Gamma_{-i}[\mathbf{N'}], \qquad (12.7)$$

where the interactions are given in the factorized form

$$\Gamma_i[\mathbf{N'}] = (1 - N_i'(\mathbf{x}))N_{-i}'(\mathbf{x})(1 - N_{-i}'(\mathbf{x} + r\mathbf{c}_i))N_i'(\mathbf{x} + r\mathbf{c}_i), \qquad (12.8)$$

and

$$\mathbf{N'} = \mathbf{N} + \Delta(\mathbf{N}). \qquad (12.9)$$

Now recall from Chapter 2.7 that the right-hand side of equation (12.6) is the sum of the α-momentum carried in the β-direction by all particle populations. The interactions (12.8) play a special role here: prior to the propagation step, they transport Γ_i units of $\mathbf{c}_i$-momentum r lattice units in the negative $\mathbf{c}_i$-direction. Thus, in equilibrium, the momentum flux in equation (12.6) must be modified such that

$$\Pi_{\alpha\beta}^{(0)} = \sum_i [N_i^{eq} - r\Gamma_i(\mathbf{N}^{eq})]c_{i\alpha}c_{i\beta}. \qquad (12.10)$$

For the moment we seek only the isotropic pressure p for a zero-velocity equilibrium. In this case, $N_i^{eq} = f$, the reduced density, and we find, by application of equation (A.33),

$$p = \frac{bc^2}{D}[f - rf^2(1 - f)^2]. \qquad (12.11)$$

This is the equation of state for a D-dimensional liquid-gas model at rest. For $r = 0$ it reduces to the usual lattice-gas equation of state. The non-standard form we have here has some very interesting implications, to which we now turn.

12.3   Liquid-gas transition

Figure 12.3 compares the predicted equation of state (12.11) to results from simulation. Three curves are shown: one for $r = 0$ (equivalent to a non-interacting gas), another for $r = 3$, and the last for $r = 8$. The $r = 8$ curve is the most interesting because it shows that two different densities may have the same pressure. Such a curve is possible for any value of r greater than the critical value $r_c \approx 5.2$, which is the *critical point* of the liquid-gas model. Below the critical point, equation (12.11) defines a unique equilibrium density. Above r_c (which may be calculated by setting $dp/df = d^2p/df^2 = 0$), two phases, one of a high density and

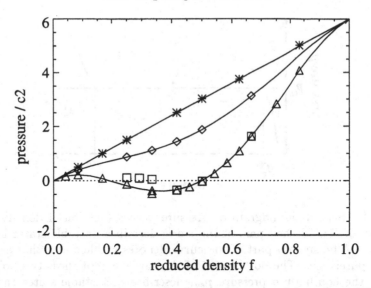

Fig. 12.3. Pressure measurements for the liquid-gas model on the FCHC lattice [12.6]. Symbols represent estimates from numerical simulations: stars are simulations of the non-interacting, ideal gas, diamonds are the liquid-gas model for $r = 3$, and triangles and squares are the liquid-gas model for $r = 8$. Triangles refer to early measurements, whereas squares have been plotted after equilibrium has been reached (except in the metastable region). Solid lines correspond to equation (12.11) in text. Pressures are divided by $c^2 = 2$.

the other a low density, may *coexist*. We call these phases *liquid* and *gas*, respectively.

An interesting theoretical question asks at what densities the liquid and gas phases co-exist for a given choice of r. One way to address this question is by the comparison of theory with simulation shown in Figure 12.3. There are two kinds of measurements in this plot. Both measurements are computations of the pressure tensor

$$\sum_i \langle n_i - r\gamma_i \rangle c_{i\alpha} c_{i\beta} \qquad (12.12)$$

for a given initial value of the reduced density f. The first measurement, represented by the stars, diamonds, and triangles, is a measure of (12.12) in the first few time steps after initialization of the liquid-gas model as a random collection of particles. We see excellent adherence to the prediction (12.11) at all points along the three curves. The second type of measurement, represented by the squares and performed only for the case $r = 8$, is more subtle. This measurement was made after waiting for the liquid-gas model to equilibrate at long times. We see the interesting consequence that some of the squares do not fall on the predicted curve. Such a departure from theory is in fact expected. To see why, we must

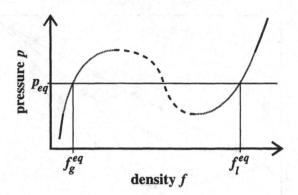

Fig. 12.4. Schematic diagram of pressure p as a function of density f for a typical fluid. The dashed part of the curve describes unstable states for which $dp/df < 0$. The smooth part of the curve denotes stable states that are purely liquid or purely gas. The dotted part of the curve, except those two points that intersect the equilibrium pressure p_{eq}, describes metastable states that decay relatively slowly to equilibrium. The equilibrium liquid and gas densities f_l^{eq} and f_g^{eq} are given by the intersection of $p(f)$ with the equilibrium pressure p_{eq}.

first recall some aspects of the thermodynamics of liquid-gas mixtures.

In a real fluid, one has an equation of state of the form $p = p(V, T)$, where V is volume and T is temperature. In our liquid-gas model, the density f is proportional to $1/V$ and the role of temperature is played by the interaction range r. In both systems the equilibrium values of p and V are stable only if $dp/dV \leq 0$. Otherwise a small increase in V due to a fluctuation would not be counteracted by a restoring force and there would be uncontrolled growth of V. In terms of the equation of state (12.11), this means that the parts of the curve where $dp/df < 0$ are unstable, and therefore inconsistent with equilibrium. Instead, if a system is initialized in this unstable region, we expect it to spontaneously separate into stable liquid and gas phases. The two phases coexist at a unique equilibrium pressure p_{eq} and corresponding equilibrium liquid and gas densities f_l^{eq} and f_g^{eq}. The equilibrium pressure is indicated by the three squares plotted for unstable values of f in Figure 12.3 and by the horizontal line in the schematic diagram shown in Figure 12.4. From the conservation of mass we have

$$f = f_l^{eq}\phi_l + f_g^{eq}(1 - \phi_l) \tag{12.13}$$

where ϕ_l, the volume fraction occupied by the liquid phase in equilibrium, is determined by the choice of f. The "equilibrium" pressures measured outside the unstable region but at densities insufficiently high to define a unique pressure are in fact metastable rather than stable, since they do not correspond to the equilibrium pressure p_{eq}.

The most interesting behavior of the liquid-gas model is its non-equilibrium decay to the equilibrium state, i. e., the dynamical evolution of phase separation. If the model is initialized in the unstable region, then its resulting behavior is qualitatively similar to the ILG phase separation shown in Figures 9.2, 11.3, and 11.4. One significant difference, however, is that whereas the volume fraction of, say, the red phase in the ILG is fixed by the red concentration in the initial condition, in the liquid-gas model the relative size of each region depends on the initial total density of the single species. Thus a particularly interesting form of phase separation can occur if the particles slowly evaporate as the liquid and gas domains grow in size. In this case, as a consequence of equation (12.13), the liquid region will occupy a progressively smaller volume fraction as time proceeds. An example of such a simulation is shown in Figure 12.5. The resulting structures are similar to those seen in two-dimensional soap films.

12.4 Hydrodynamics

Given our previous developments in Chapters 2 and 3, it is straightforward to derive an inviscid hydrodynamic equation for the liquid-gas model. We begin, as usual, with the momentum-balance equation

$$\partial_t(\rho u_\alpha) = -\partial_\beta \Pi^{(0)}_{\alpha\beta}. \tag{12.14}$$

Then, using the same manipulations we performed in Chapter 3.5 we substitute equation (12.10) for the momentum flux density tensor $\Pi^{(0)}_{\alpha\beta}$ to obtain the Euler equation

$$\partial_t \rho u_\alpha + \partial_\beta[g_r(\rho)\rho u_\alpha u_\beta] = -\partial_\alpha[p_r(\rho, u^2)]. \tag{12.15}$$

This equation has the same form as the Euler equation (3.47) of the non-interacting lattice gas. There are some key differences, however, and they are each parameterized by the interaction range r. We give the results below for D-dimensional lattices. The factor $g(\rho)$ is now

$$g_r(\rho) = \frac{D}{D+2} \frac{1-2f}{1-f} \left(1 + \frac{4rf^2(1-f)^2}{1-2f}\right). \tag{12.16}$$

The equation of state for a uniform equilibrium with velocity **u** is

$$p_r(\rho, u^2) = c_{s0}^2 \rho - \frac{br}{D} f^2(1-f)^2 + \frac{1}{2}\rho \left(g_r(\rho) - g_0(\rho)c_{sr}^2(D+2)\right) u^2, \tag{12.17}$$

in which the square of the soundspeed, $c_{sr}^2 = \mathrm{d}p_r/\mathrm{d}\rho$, is given by

$$c_{sr}^2(\rho) = \frac{1}{D}[1 - 2rf(1-f)(1-2f)]. \tag{12.18}$$

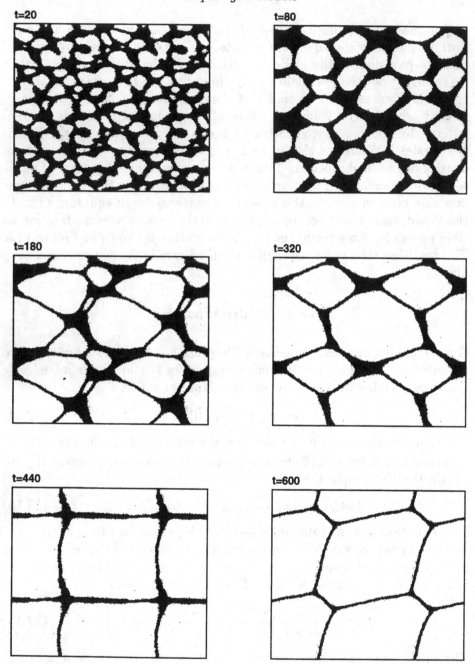

Fig. 12.5.   Phase separation with particle removal in the 2D liquid-gas model
[12.3]. The pixels are black when there are more than two particles per site
and white otherwise. Thus the liquid phase is mostly black while the gas phase
is mostly white. The lattice is initialized with a uniform particle distribution.
As time progresses particles are slowly removed at random. After an initial
transient, the density of the liquid and the gas remain constant, but the fraction
of space covered by the dense phase is decreasing. This leads to the formation
of a 2D soap froth.

Once again, at $r = 0$ one recovers the results for a non-interacting lattice gas. The most significant difference for $r > 0$ is of course the equation of state (12.17), for which we have already discussed the ramifications. Another important difference is in the factor $g_r(\rho)$. Comparing it to the 2D expression (3.48) for a non-interacting gas, we see that it contains explicit effects of the interactions. These are due to the fact that i) the rate of interaction depends on the distributions of particles and holes in the b directions of the lattice; and ii) momentum transfer depends on the rate of interactions. Depending on the velocity and density there may be more or less particles and holes available for interaction in a given direction, and thus more or less momentum transfer.

Lastly we discuss the viscous behavior of the liquid-gas model. Here there is also a significant change from the behavior of the non-interacting gas. To see it, consider once again the schematic view of molecular motion in a shear flow $\partial_y u_x$ shown in Figure 2.4. In the relevant discussion in Chapter 2.5, we showed how viscous stress is proportional to $\partial_y u_x$, with the constant of proportionality given by the viscosity ν. The viscous drag results from molecular mixing in the shear direction. The Maxwell estimate (4.33) gives $\nu \sim \ell_{\mathrm{mfp}} U$, where ℓ_{mfp} is the mean free path and U is the "thermal velocity." For the non-interacting lattice gas we have $U = |\mathbf{c}|$ lattice units per time step. Interactions, however, carry momentum over distances of order r at a speed U which is also of order r. Thus we find that $\nu \sim r^2$.

12.5 Exercises

12.1 Rederive the equation of state (12.17) for a model in which interactions at a specified range r occur at a rate $\mu < 1$.

12.6 Notes

The liquid-gas model was introduced in:

12.1 Appert, C. and Zaleski, S. (1990). A lattice gas with a liquid-gas transition. *Phys. Rev. Lett.* **64**, 1–4.

It was studied further in:

12.2 Appert, C., Rothman, D. H., and Zaleski, S. (1991). A liquid-gas model on a lattice. *Physica D* **47**, 85–96.

12.3 Appert, C. and Zaleski, S. (1993). Dynamical liquid-gas phase transition. *J. Physique II France* **3**, 309–337.

These papers describe models that are somewhat more complicated than the model of this chapter, in that they allow for more types of collisions. The "minimal model" that we presented here was introduced in:

12.4 Appert, C., d'Humières, D., and Zaleski, S. (1993). Gaz sur reseau avec interactions attractives minimales. *C. R. Acad. Sci. Paris* **316**, 569–574.

The minimal model allows for the easy extension to 3D that we described. It is studied in further detail in:

12.5 Appert, C., d'Humières, D., Pot, V., and Zaleski, S. (1993). Three dimensional lattice gas with minimal interactions. *Transport Theory and Statistical Physics* **23**, 107–122.

12.6 Appert, C., Pot, V., and Zaleski, S. (1996). Liquid-gas models on 2D and 3D lattices. In Lawniczak, A. and Kapral, R., editors, *Pattern Formation and Lattice Gas Automata, Fields Institute Communications* **6**, pages 1–12 (American Mathematical Society, Providence).

The above papers form part of:

12.7 Appert, C. (1993). *Transition de phase dynamique de type liquide-gaz et création d'interfaces dans un gaz sur réseau.* Ph. D. thesis, Université Pierre et Marie Curie (Paris VI).

Interfacial properties of the liquid-gas model are studied in:

12.8 Appert, C. and d'Humières, D. (1995). Density profiles in a diphasic lattice-gas model. *Phys. Rev. E* **51**, 4335–4345.

Viscous properties of the model are investigated in:

12.9 Pot, V., Appert, C., Melayah, A., Rothman, D. H., and Zaleski, S. (1996). Interacting lattice gas automaton study of liquid-gas properties in porous media. *J. Physique II France* **6**, 1517–1534.

This paper also includes a study of interactions with walls and applications to evaporation in porous media.

Elementary aspects of the liquid-gas transition are discussed in:

12.10 Reif, F. (1965). *Fundamentals of statistical and thermal physics* (McGraw-Hill, New York).

13
Flow through porous media

We now take a break from the theoretical developments of the previous chapters and consider applications of lattice gases to the study of complex flows through complex geometries. Our complex geometry is one of the most complicated nature has to offer—a porous rock. The flows we consider are either those of a simple fluid, such as water, or an immiscible two-fluid mixture, such as water and oil. The problems we shall illustrate are not only of intrinsic interest for physics but have applications in fields as diverse as hydrology, oil recovery, and biology, to name just a few.

Our objectives in this chapter are twofold. First, we wish to indicate the level of accuracy that one may expect from these kinds of flow simulations. Second, we wish to show what we can learn from such work. We begin with a brief introduction to the subject.

13.1 Geometric complexity

All rocks found on the earth's surface are porous, but those rocks that we call *sandstones* are usually more porous than others. Sandstones are formed from random assemblages of sand grains that are cemented together over geologic time. Fluids such as oil or water may then become trapped in the pore space between the cemented sand grains. There may be an economic interest in extracting these fluids, or, equally possible in modern times, we may wish to predict the rate at which some contaminant such as radioactive waste could migrate through such a medium. Thus in both cases an understanding of how the geometry of a porous rock influences flow through it is of crucial importance.

Probably the best way to consider the geometric complexity of a real sandstone is to look at it. Figure 13.1 is a two-dimensional slice through a 3D tomographic reconstruction of a small piece of *Fontainebleau sandstone*, and Figure 13.2 is a 3D image. Although Fontainebleau sandstone

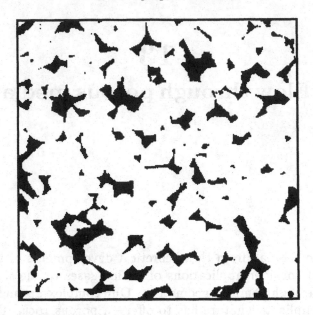

Fig. 13.1. 2D slice through a 3D tomographic reconstruction of a small piece of Fontainebleau sandstone [13.7]. The pore (void) space is indicated by black and the linear dimension is 2.2 mm. There are 288 pixels on a side and the resolution is 7.5 μm. The tomographic technique employs an X-ray synchrotron source to allow the high-resolution imaging.

is one of the cleanest sandstones available (it consists mostly of pure quartz sand grains without much variation in size), this particular type of rock still presents considerable disorder. We wish to understand how fluids flow through the pore space, which occupies a volume fraction of about 15% of this rock. Obviously any flow through this structure will be rather complicated.

Figure 13.3 shows what one such flow looks like. In this case the simulation was performed using a lattice-Boltzmann method. The main thing to notice is that most of the flow occurs in a small fraction of the available pore space. This is because most of the possible flow paths are blocked by some kind of obstacle, and the resulting flow is effectively "focused" on the least resistant of the remaining paths. The effect would be even more exaggerated if the system were bigger, but would be harder to display graphically.

Our principal concern is to show how we may use our ability to perform simulations such as these so that we may learn something new about flow through porous media. First, however, we must review some of the basic ideas in the subject.

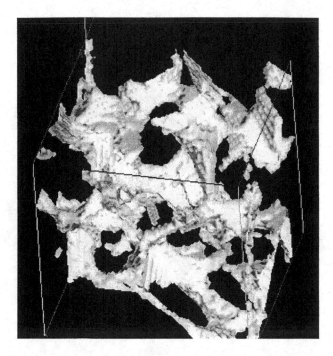

Fig. 13.2. 3D reconstruction of the pore space (shown in gray) of Fontainebleau sandstone [13.7]. The linear scale is 64 voxels, or 0.5 mm, on a side.

13.2 Another macroscopic scale

The science of flow through porous media had its genesis in the mid-nineteenth century work of the French engineer Henri Darcy. By conducting experiments with sand packs, Darcy found that the flow rate **J** of fluid flowing through porous media was proportional to the force applied to it. In symbols, we write *Darcy's law* as

$$\mathbf{J} = -\frac{k}{\mu}(\nabla p - \rho \mathbf{g}), \tag{13.1}$$

where μ is the viscosity of the fluid, ∇p is the pressure gradient applied to it, $\rho\mathbf{g}$ is the gravitational force density, and k is the *permeability*. The permeability coefficient has dimensions of area, and is a measure of the conductivity of the porous medium to fluid flow through it.

Of the many possible sources of interest, we wish to emphasize three interesting aspects of Darcy's law (13.1). The first is purely a question of mathematical physics. For an incompressible fluid, we may be reasonably confident that the flow at the scale of a pore is described by the Navier-Stokes equations (2.8) and (2.32). The question, briefly put, is how the linear Darcy equation (13.1) can result from the nonlinear Navier-Stokes

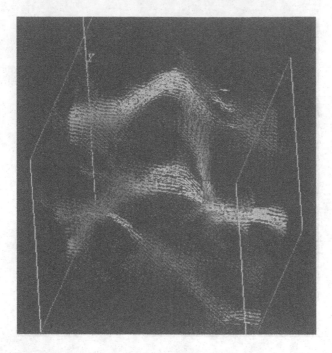

Fig. 13.3. Flow through the sandstone shown in Figure 13.2 [13.6]. Bright shades of gray indicate fast flow speeds. There is no flow in the center because it is occupied by a sand grain.

equation (2.32) at a larger scale L_{darcy} for which material properties such as permeability and porosity (i.e., the void fraction) vary slowly in space. Indeed, the linearity of Darcy's law is hardly obvious because it is known to be applicable to flows in which the nonlinear term of the Navier-Stokes equation (2.32) plays a role of comparable magnitude to the viscous term.

The second cause for interest in Darcy's law derives from the permeability coefficient. Despite the existence of many empirical scaling laws, there is still no theoretical procedure—other than numerical integration of the Navier-Stokes equations—for the accurate estimation of k from knowledge of the microgeometry. Indeed, commonly employed scaling laws may be in error by an order of magnitude or more.

Lastly, and at least as importantly, one would like to know whether there exists a constitutive equation or system of equations similar to Darcy's law that applies to the flow of two or more immiscible fluids in porous media. Indeed, it is commonly assumed that a simple modification of equation (13.1) applies to multiphase flow. For two fluids, such an equation usually takes a form similar to

$$\mathbf{J}_i = k_i^{(r)}(S)\frac{k}{\mu_i}\mathbf{X}_i, \qquad i = 1, 2, \tag{13.2}$$

where S is the relative saturation of one of the fluids, μ_i and $k_i^{(r)}(S)$ are the viscosity and the saturation-dependent *relative permeability* coefficient, respectively, for the ith fluid, and $\mathbf{X}_i$ is the body force (for example, $\mathbf{X}_i = \rho_i \mathbf{g}$) acting on the ith fluid. Among the many possible problems with equation (13.2) is its implicit assumption that the flow of each fluid is uncoupled to the other; i.e., that each fluid flows through a porous medium composed of the real porous medium plus the other fluid, with no viscous interaction between the two fluids. Whether this assumption has any validity is one question that may be addressed by numerical simulations of flow through geometries such as those shown in Figure 13.2.

In this chapter, we show how one can answer such questions by simulations of lattice-gas (and related lattice-Boltzmann) models. These methods are useful for flow through porous media because they are easy to apply to problems with irregular geometries, particularly in the case of two-phase flow. This raises a curious point, however. In our theoretical studies of lattice gases, our principal concern has been the derivation of macroscopic hydrodynamic equations from microscopic models. In the terminology put forth earlier in Sections 2.1 and 4.2, we have derived macroscopic equations at the scale L_{hydro} from microscopic models defined at the scale $\ell_{\text{lattice}} \sim \ell_{\text{mfp}}$. And now we have the audacity to propose measuring quantities in simulations performed at the scale $L_{\text{darcy}} \gg L_{\text{hydro}} \gg \ell_{\text{mfp}} \sim \ell_{\text{lattice}}$! Can we really do that?

The answer is yes. There are many ways to show why, but it is perhaps most interesting to consider how simulations of single-phase flow through geometries like Figure 13.2 compare to flows through the real rock itself. We do so in the following section.

13.3   Single-phase flow

Current technology dictates that neither tomographic imaging nor numerical simulation can deal with 3D images whose cross-sections are much larger than that shown in Figure 13.1. Laboratory measurements of permeability, on the other hand, are typically made at scales at least one order of magnitude larger. In asking whether results derived numerically at the small scale of Figure 13.1 have any relevance to the larger scale, we must make a practical estimate of the scale L_{darcy}.

We first describe how to compute the permeability with the lattice-gas or lattice Boltzmann method. As described in Section 7.3, one typically forces fluid through a periodic box by adding, on average, g units of momentum in the flow direction at each site at each time step. This imitates the gravity term in equation (13.1). Once the flow has reached a steady state, the permeability k is computed from equation (13.1). This gives, for a force applied in the z-direction,

$$k = \frac{\phi \nu h^2 \sum_{\mathbf{x}} \sum_i \langle n_i(\mathbf{x}) \rangle c_{iz}}{\sum_{\mathbf{x}} g(\mathbf{x})}, \qquad (13.3)$$

where the sums over $\mathbf{x}$ are performed only in the void space, ν is once again the kinematic viscosity, ϕ is the porosity, and h is the physical dimension of the lattice spacing as defined in Section 7.1. The factor of ϕ is required because $\mathbf{J}$ is defined as a volumetric flow rate through a unit area, and the factor h^2 provides the appropriate dimensional units. The average $\langle n_i \rangle$ is computed over many time steps if one uses a lattice gas or is read as N_i if one uses the Boltzmann method.

We now describe a determination of L_{darcy}. From a slab of Fontaine-bleau sandstone, we extract

- a core of diameter 2 cm and length 4 cm for laboratory measurements; and

- three smaller samples of linear dimension of about 2 mm for tomographic imaging.

One of the tomographic images is made up of 224^3 voxels, or volume elements. We subdivide it into eight "mini"-blocks of size 112^3, and we further subdivide each mini-block into eight "micro"-blocks of size 56^3. Then we use the lattice-Boltzmann method to compute the permeability in all 64 micro-blocks, all eight mini-blocks, and the original large sample of size 224^3. Lastly, as a crude measure of the dependence of k on the tomographic sample and the computational method, we compute the permeability of the other two large tomographic images using a classical finite-difference solution to the Stokes equations.

The results are shown in Figure 13.4, where we have plotted the permeability versus the porosity for all 75 mm-scale samples in addition to the permeability and porosity obtained by laboratory measurement of the cm-scale core. (Porosity is obtained simply by dividing the number of void sites by the sample size.) There are several interesting remarks that may be made. First, we see that the porosity ϕ varies over approximately a factor of four for the smallest samples, over about 10% for the intermediate samples, and insignificantly for the largest samples. Thus this purely geometric quantity seems to vary slowly at the intermediate scale. Second, we see that permeability scales roughly like ϕ^3, in accord with many empirical scaling laws. Third—and most interesting—is the size-dependence in the scatter of the permeability. This scatter ranges over nearly *three* orders of magnitude for the smallest samples, by a factor of two for the intermediate samples, and by about 30% for the three large

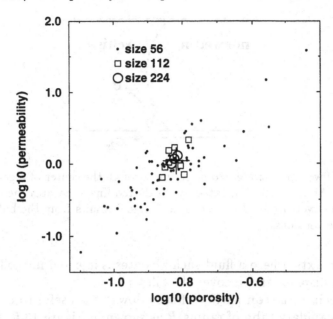

Fig. 13.4. Permeabilities and porosities calculated from tomographic reconstructions of Fontainebleau sandstone [13.7]. The sizes 56, 112, and 224 correspond to the linear dimension (in voxels of size 7.5 μm) of the sample used for the computation, and correspond, roughly, to 0.4, 0.8, and 1.7 mm, respectively. The permeability of two of the 1.7 mm samples was computed by a classical finite-difference method, while all other computations were with the lattice-Boltzmann method. The permeability and porosity measured experimentally from a cm-scale core of Fontainebleau sandstone is given by the center of the symbol +.

samples. Thus the scale L_{darcy} where both permeability and porosity vary slowly in space appears to be about the size of our largest sample.

The last item of interest in Figure 13.4 is the comparison of computations to laboratory measurements. One sees that the lattice-Boltzmann computation of permeability for the largest sample compares well not only to the finite-difference computations but also to the laboratory measurement for the cm-scale core.

13.4 Two-phase flow: experiment versus simulation

Although single-phase flow through porous media is by itself somewhat complicated, its complexity is only a fraction of that of two-phase flow. Two-phase flow through rocks occurs naturally as oil migrates upward from source to reservoir rocks and as water flows through partially saturated soils. It is also of considerable engineering importance in oil pro-

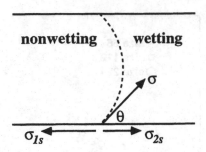

Fig. 13.5. Two-dimensional cross-section through the center of a capillary tube of radius R. An interface, indicated by the dotted line, separates the non-wetting fluid from the wetting fluid. The contact angle θ results from the balance of the three surface tensions.

duction contexts, where a fluid such as water is injected into oil reservoirs in order to displace and recover the oil.

To appreciate the complexity of such flows, it is useful to consider two fluids in a capillary tube of radius R as shown in Figure 13.5. One of the fluids, called the *wetting* fluid, spreads preferentially over solid surfaces. The other fluid is called *non-wetting*. The preferential spreading of the wetting fluid results from stronger attractive molecular forces between it and the solid surface as compared to that of the non-wetting fluid and the solid surface. Since there are three different interfaces, there are three different surface tensions: the fluid-fluid surface tension σ and the two fluid-solid surface tensions σ_{1s} and σ_{2s}. For the interfaces to be in thermodynamic equilibrium, the three surface tensions must balance such that

$$\sigma \cos \theta = \sigma_{1s} - \sigma_{2s}, \qquad (13.4)$$

where fluid 1 is taken to be non-wetting. Equation (13.4), known as *Young's law*, defines the contact angle θ. A little geometry shows that the radius of curvature of the meniscus is $R/\cos\theta$. From Laplace's law (11.16) we can then calculate the pressure difference $(\Delta p)_0 = p_1 - p_2$ for an interface at rest. We find

$$(\Delta p)_0 = \frac{2\sigma \cos \theta}{R}. \qquad (13.5)$$

If the interface is to move, however, the pressure drop must differ from $(\Delta p)_0$. Under the appropriate boundary conditions the non-wetting fluid would penetrate the tube if $\Delta p > (\Delta p)_0$, whereas the wetting phase would be imbibed into the tube in the opposite case.

The capillary forces acting in two-phase flows thus severely complicate what is a relatively straightforward problem for the case of a single fluid. Nevertheless, typical engineering practice assumes that equation (13.2), the two-phase extension of Darcy's law, describes the simultaneous flow of two immiscible fluids at the scale L_{darcy}. Although this empirical law has survived since its inception decades ago, its ongoing presence owes more to a lack of understanding rather than any theoretical justification. Thus a systematic exploration of its validity can offer valuable insight.

We first consider a laboratory experiment and compare it to simulation [13.7]. In the experiment, a non-wetting fluid (oil) is injected into a sample of Fontainebleau sandstone that is initially filled with a wetting fluid (water). The oil is injected until no more water may be expelled from the sample. The fraction of the pore space occupied by the water is then the residual water saturation S_{rw}. A laboratory measurement of $k_1^{(r)}(S_{\mathrm{rw}})$, the relative permeability of the rock to the non-wetting fluid (fluid 1), is then obtained by measuring the flux of oil, assuming a linear response, and using equation (13.2). The measurements found $S_{\mathrm{rw}} = 0.03$ and $k_1^{(r)}(S_{\mathrm{rw}}) = 0.92$

One way to estimate $k_1^{(r)}(S_{\mathrm{rw}})$ by simulation is simply to imitate as closely as possible the laboratory experiment. Such a simulation is shown in Figure 13.6. Here a non-wetting fluid is injected into one face of a 3D sample of size $104 \times 104 \times 208$ that is initially filled with wetting fluid. The simulation shown was performed with the 3D immiscible lattice gas. The boundaries parallel to the flow direction are jacketed by solid walls but the boundaries perpendicular to the flow are periodic to the particle occupancy variable n_i. Injection of the non-wetting red fluid is simulated by imposing that the color of all incoming particles at the top boundary be red while requiring that all incoming particles on the bottom boundary be blue. The relative wettability of the two fluids is controlled by assigning a color to solid sites. In the case shown, the solid sites were colored blue and thus the blue fluid is wetting.*

The saturation of the wetting fluid was monitored during the simulation. When it decreased to a steady state, the relative permeability coefficient $k_1^{(r)}(S_{\mathrm{rw}})$ was computed. If we assume a linear response and once again take the non-wetting fluid to be red and the force to be in the z-direction, the simplest way to compute $k_1^{(r)}(S_{\mathrm{rw}})$ is to calculate the saturation-dependent total momentum of the red phase,

$$Q(S) = \sum_{\mathbf{x}} \sum_i \langle r_i(\mathbf{x}) \rangle c_{iz}, \tag{13.6}$$

---

* The relative wettability of 3D ILG fluids is controlled by biasing the color gradient (11.3). Specifically, when the gradient $\mathbf{f}$ is calculated at sites neighboring solid sites, the quantity $\sum_i (r_i - b_i)$ at the solid sites is set to an integer between -24 and 24. As the magnitude of this integer decreases towards zero, the contact angle increases toward 90 degrees, the condition of neutral wettability.

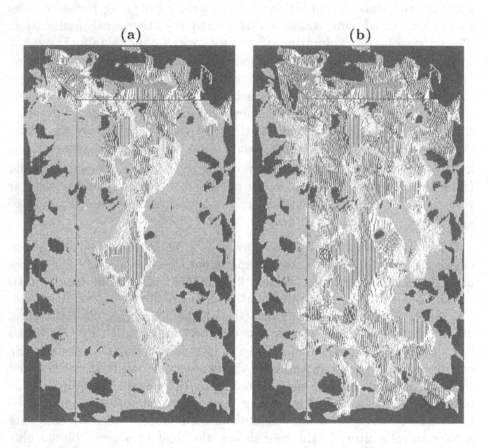

Fig. 13.6. Simulation of injection of nonwetting fluid into a sample of Fontainebleau sandstone that is initially filled with wetting fluid [13.7]. The sample size is $104 \times 104 \times 208$. The invading fluid enters the sample at the top; the sample is viewed from above and to the left. The background is black, the solid porous matrix is transparent and colorless, the wetting fluid is transparent and grey, and the nonwetting fluid is opaque and white. Thus black also indicates that the solid matrix is contiguous in the line of sight. The left-hand panel (a) shows the "breakthrough" configuration, i.e., just after the non-wetting fluid penetrates the full extent of the medium. The right-hand panel (b) shows the configuration near steady state. Because the sample is jacketed by solid walls, which are themselves covered by a thin layer of wetting fluid, the left and right edges of the image are dominated by the grey wetting fluid. However, these edges correspond to a small fraction of the total volume.

where the sums are performed in steady state for a wetting-fluid saturation S. Using equation (13.2), we find that the relevant relative permeability coefficient is then

$$k_1^{(r)}(S_{\rm rw}) = \frac{Q(S_{\rm rw})}{Q(0)} \qquad (13.7)$$

where $Q(0)$, the total momentum in a single-phase flow simulation, is computed for the same level of forcing. The simulation found $k_1^{(r)}(S_{\rm rw}) = 0.84$, in reasonable agreement with the laboratory measurement, but the calculated residual saturation, $S_{\rm rw} = 0.53$, was strongly overestimated. While there may be many possible sources of such an error, the most likely culprit is the poor spatial resolution of the simulation. The problem appears to be that the small size of the reconstructed pores, on the order of just a few lattice units, requires that wetting layers be much thicker (i.e., about one lattice unit) relative to the pore size than one expects in real experiments. Indeed, for the tomographic sample studied, 54% of the void sites are adjacent to solid sites. Thus the computed residual saturation of 53% could be consistent with drainage of nearly all wetting fluid except that adjacent to pore walls, and is thus not so obviously inconsistent with the experimental value of 3%.

A second type of numerical simulation allows a calculation of the relative permeability $k_1^{(r)}(S_{\rm rw})$ that is much simpler [13.7]. In this case the pore space is initially filled with a homogeneous mixture of wetting and non-wetting fluids with wetting saturation chosen to match the value obtained in the laboratory experiment: $S_{\rm rw} = 0.03$. The mixture is then forced through the rock with fully periodic boundary conditions in the flow direction so that the saturation may be kept constant. Only the nonwetting fluid is forced, and the two fluids undergo phase separation as they flow. The relative permeability $k_1^{(r)}(S_{\rm rw})$ is then measured after the flow reaches steady state. Such simulations found $k_1^{(r)}(S_{\rm rw}) = 0.90 \pm 0.02$, in good accord with the laboratory experiment.

13.5 Two-phase flow: theory versus simulation

Although the previous section summarized some interesting comparisons between simulation and experiment, we believe that a more valuable use of such simulations is to explore the validity of empirical theories such as the two-fluid Darcy equation (13.2). In other words, once we are satisfied that our models simulate the relevant physical processes, we ask not whether they can reproduce the theoretical prediction of the Darcy equation, but whether the Darcy equation is itself correct.

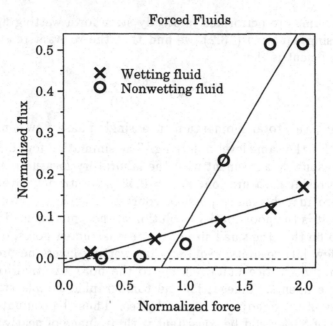

Fig. 13.7. Flux as a function of force in simulations of two-phase flow through Fontainebleau sandstone [13.8]. The saturation $S = 0.5$ for all data points. The symbol × ·represents the response of the wetting fluid when only the wetting fluid is forced, while the symbox ○ represents the response of the non-wetting fluid when only the non-wetting fluid is forced. The straight line shows the linear trend, which exists only for fast fluxes in the non-wetting case. Error bars are approximately the size of the symbols.

We consider a generalization of equation (13.2) that includes the possibility for viscous coupling between the two fluids. We write

$$\mathbf{J}_i = \sum_j L_{ij}(S)\mathbf{X}_j, \qquad i,j = 1,2, \tag{13.8}$$

where, for the case of equal viscosities considered here, $L_{ij} = \ell_{ij}(\theta)k/\mu$, with $0 \le \ell_{ij} \le 1$. The ℓ_{ij} are generalized relative permeability coefficients; the cross terms ℓ_{12} and ℓ_{21} reflect momentum transfer from one fluid to the other.

Equation (13.8) is the most general form of a linear response law for the Darcy-scale flow of two fluids through porous media. As such, it takes the form of a coupled force-flux relation that one finds in the theory of irreversible thermodynamics. In that context, the $\mathbf{J}_i$'s and the $\mathbf{X}_j$'s are known as conjugated fluxes and forces, respectively, entropy production is given by $\sum_i \mathbf{J}_i \cdot \mathbf{X}_i$, and *Onsager reciprocity* predicts that $L_{12} = L_{21}$. However, none of these theoretical predictions apply *a priori* to two-phase flow through porous media. We therefore describe a series of simulations

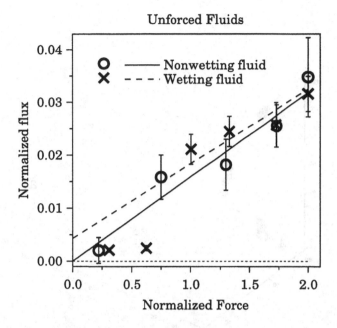

Fig. 13.8. Flux as a function of force in simulations of two-phase flow through Fontainebleau sandstone [13.8]. The saturation $S = 0.5$ for all data points. Unlike Figure 13.7, the data points now represent the response of the *unforced* fluids. Thus the symbol × represents the response of the wetting fluid when only the non-wetting fluid is forced, while the symbox ○ represents the response of the non-wetting fluid when only the wetting fluid is forced. The straight lines are least-squares fits to the linear trends.

that attempt to respond to the following questions:

- Is the hypothesized linear response (13.8) correct?

- Are the cross terms L_{12} and L_{21} significant?

- Are the cross terms symmetric; i.e., does $L_{12} = L_{21}$?

We first examine the question of linearity. As in the second simulation described in the previous section, we simulate flows in a tomographic reconstruction of Fontainebleau sandstone for which the boundary conditions are fully periodic and the saturation S is kept constant. Figure 13.7 shows results from two numerical experiments: one in which only the nonwetting fluid is forced, and another in which only the wetting fluid is forced. In both cases the saturation $S = 0.5$ and the data points show the flux of the *forced* fluids only. We see that the wetting fluid exhibits an approximately linear response but the non-wetting fluid clearly responds nonlinearly. This nonlinear behavior is the macroscopic manifestation of

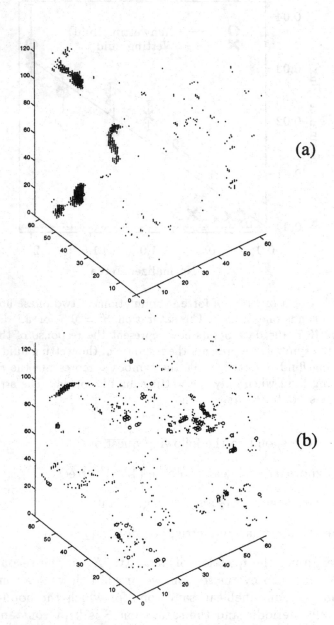

Fig. 13.9. Vertical velocities in simulated flow through Fontainebleau sandstone for (a) a two-phase mixture of saturation $S = 0.5$ in which only the non-wetting fluid is forced and (b) the same mixture when only the wetting fluid is forced [13.8]. The plots show the locations in space that are among the fastest one percent (dots) and fastest tenth of a percent (circles) of all sites. In (a) most of the circles overlap too much to be visible.

the capillary forces associated with three-phase contact lines such as that pictured in Figure 13.5. Once the flow is fast enough, however, the effects of capillarity are less significant and the data show that an approximately linear response may be valid. If we fit both linear regimes to equation (13.8), we find that the non-wetting relative permeability $\ell_{11} \approx 0.5$ and that the wetting relative permeability $\ell_{22} \approx 0.1$. The lower relative permeability of the wetting phase results because it creeps along the pore walls rather than flowing in the center of channels like the non-wetting phase.

The same pair of numerical experiments allows us to address the remaining questions concerning the cross terms. Figure 13.8 shows the response of the *unforced* fluids. We see that the responses are not only approximately linear but also approximately symmetric. Specifically, least-squares fits to the data yield $\ell_{12} = 0.015 \pm 0.003$ and $\ell_{21} = 0.014 \pm 0.004$. Though the size of the cross terms is not large, one sees from comparison of Figure 13.7 and 13.8 that the fluxes of the forced fluids are only about one order of magnitude larger than that of the unforced fluids.

Thus we may tentatively conclude that the generalized linear response relation (13.8) has only a limited regime of applicability. In that regime it appears that the cross terms may be symmetric and of significant magnitude. Why would the cross terms be symmetric in such far-from-equilibrium flows? We don't have an answer, but it is nonetheless interesting and useful to look at what these flows actually look like. Figure 13.9 shows perspective views of the velocity field inside the three dimensional porous rock. The two plots shown correspond to two-phase flow in which only the non-wetting fluid is forced and two-phase flow in which only the wetting fluid is forced. In the former case, one sees that most of the flow occurs as a well-organized narrow tortuous channel. In the latter case, however, we see little or no organization of the flow. Thus two highly disparate flows appear to result in equal transfers of momentum from the forced to the unforced phases. How this could be so, and whether it is indeed a general result, remains to be adequately addressed.

13.6 Notes

In the interest of brevity we have not included many of the numerical validations available in the literature on lattice-gas and lattice-Boltzmann simulations of flow through porous media. Perhaps the most important of these validations is that Darcy's law indeed holds for flows through complex structure. This was shown first in:

13.1 Rothman, D. H. (1988). Cellular-automaton fluids: A model for flow in porous media. *Geophysics* **53**, 509–518.

Comparisons of calculated permeability to theoretical predictions for flows through assemblages of spheres are given in:

13.2 Cancelliere, A., Chang, C., Foti, E., Rothman, D. H., and Succi, S. (1990). The permeability of a random medium: Comparison of simulation with theory. *Physics of Fluids A* **2**, 2085-2088.

Aspects of single-phase flow are also discussed in:

13.3 Chen, S., Diemer, K., Doolen, G., Eggert, K., Fu, C., Gutman, S., and Travis, B. J. (1991). Lattice gas automata for flow through porous media. *Physica D* **47**, 72–84.

Validations of two-phase flow may be found in:

13.4 Rothman, D. H. (1990). Macroscopic laws for immiscible two-phase flow in porous media: Results from numerical experiments. *J. Geophys. Res.*, 95:8663.

13.5 Gunstensen, A. K. and Rothman, D. H. (1993). Lattice-Boltzmann studies of two-phase flow through porous media. *J. Geophys. Res.* **98**, 6431–6441.

The first-of these papers includes a validation of Young's law (13.4) in 2D. The second paper includes an explicit test of the relative permeability equations (13.8) for the case of annular flow in a cylinder, for which exact expressions may be derived for the coefficients $L_{ij}(S)$.

The precise location of boundary conditions discussed in Section 7.4 is of crucial importance in flow through porous media. A practical example of how permeability can depend on viscosity in simulations of the single-time relaxation-Boltzmann model is given in:

13.6 Ferréol, B. and Rothman, D. H. (1995). Lattice-Boltzmann simulations of flow through Fontainebleau sandstone. *Transport in Porous Media* **20**, 3–20.

The comparison of simulations with laboratory measurements is detailed in:

13.7 Auzerais, F. M., Dunsmuir, J., Ferréol, B., Martys, N., Olson, J. F., Rama-krishnan, T. S., Rothman, D. H., and Schwartz, L. M. (1996). Transport in sandstone: a study based on three-dimensional microtomography. *Geophys. Res. Lett.* **23**, 705–708.

A more extensive study of two-phase flow itself is given in:

13.8 Olson, J. F. and Rothman, D. H. (1996). Two-fluid flow in sedimentary rock: simulation, transport, and complexity. *J. Fluid Mech.*, in press.

These results, which include the study of Onsager reciprocity, also form part of Olson's thesis [11.3].

The liquid-gas model of Chapter 12 may be applied to the study of evaporation in porous media. An exploration of this idea, including validations of the liquid-gas model itself, is in:

13.9 Pot, V. (1994). *Étude microscopique du transport et du changement de phase en milieu poreux, par la méthode des gaz sur réseau*. Ph. D. thesis, Université Pierre et Marie Curie (Paris VI).

Aspects of this work may also be found in Ref. [12.9].

A review of many of the ways in which numerical simulation may be applied to flow through porous media, by both lattice-gas and finite-difference methods, is given in:

13.10 van Genabeek, O. and Rothman, D. H. (1996). Macroscopic manifestations of microscopic flows through porous media: phenomenology from simulation. *Annual Reviews of Earth Sciences* **24**, 63–87.

Many references to the subject beyond those given here may be found in this paper.

Suspensions of hard spheres may be thought of as unconsolidated porous media. A detailed presentation of how the lattice-Boltzmann method may be applied to such problems is given in:

13.11 Ladd, A. J. C. (1994). Numerical simulations of particulate suspensions via a discretized Boltzmann equation. Part 1. Theoretical foundation. *J. Fluid Mech.* **271**, 285–310.

13.12 Ladd, A. J. C. (1994). Numerical simulations of particulate suspensions via a discretized Boltzmann equation. Part 2. Numerical results. *J. Fluid Mech.* **271**, 311–340.

14

Equilibrium statistical mechanics

In this chapter we examine the foundations of lattice-gas theory. This theory rests principally on the statistical physics of discrete systems, which in turn is founded on Markov process theory. We explain how lattice gases are probabilistic cellular automata, a particular case of Markov processes in a finite configuration space. This leads us to fundamental concepts such as the *H-theorem*. This famous theorem proves that a function identified with the negative entropy decreases or stays constant in time. This theorem is the foundation of the oft-remarked fact that lattice-gas models are in some sense "absolutely stable" since they are driven to equilibrium by the growth of entropy. This H-theorem is actually the first of a series that continues in Chapter 15. We also introduce the *Liouville equation*, and the *Gibbs equilibrium*, that was postulated in Chapter 3.

14.1 Automata

Automata are systems with a finite set of configurations X evolving deterministically from one configuration s in X to another. Thus there is a function $\mathcal{F}$ giving the time evolution

$$s(t+1) = \mathcal{F}[s(t)].\tag{14.1}$$

Automata may be described by *de Bruijn diagrams* where each action of $\mathcal{F}$ is represented by an arrow. The automaton is *invertible* when the function $\mathcal{F}$ is invertible. An example of such a diagram is shown in Figure 14.1.

A deterministic lattice gas on a domain of large but finite extent $\mathcal{L}$ is a very large automaton which is best described as a *cellular automaton*. The configurations s of a cellular automaton are the collection of cell configurations $s(t) = (s_i(\mathbf{x}, t))_{i\mathbf{x}}$. We also write $s(\mathbf{x}, t) = (s_i(\mathbf{x}), t)_i$ for the local configuration of the automaton, but we sometimes keep the notation $\mathbf{n}$ to describe a local configuration. The dynamics of the cellular

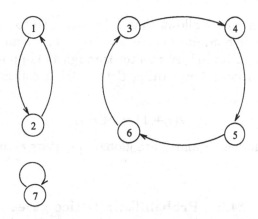

Fig. 14.1. A de Bruijn diagram for an invertible deterministic automaton. There is a finite number of configurations, and each configuration has a single successor. In an invertible automaton, the trajectories may only form closed loops, as in this example.

automaton are described by a function $\mathcal{F}$ of the local configuration $s(\mathbf{x}, t)$ and of the configurations of a neighborhood $V_{\mathbf{x}}$ of $\mathbf{x}$. In a lattice gas, neighborhoods are made of the points $\mathbf{x} + \mathbf{c}_i$. The evolution is given by

$$s(\mathbf{x}, t + 1) = \mathcal{F}(s(\mathbf{y}_1, t), \cdots, s(\mathbf{y}_n, t))_{\mathbf{y}_1, \cdots, \mathbf{y}_n \in V_{\mathbf{x}}}. \qquad (14.2)$$

Deterministic lattice gases are a particular case of cellular automata, but they have a much more natural, two step, representation than equation (14.2). First, the collision operator $\mathcal{C}$ is a local operator that maps a local configuration on the configuration after collision

$$s'(\mathbf{x}, t) = \mathcal{C}[s(\mathbf{x}, t)], \qquad (14.3)$$

where $s(t)$ is the pre-collision configuration and $s'(t)$ is the post-collision configuration. Comparing to the operator Δ in Eq. (2.33), we see that $\mathcal{C} = \mathbf{I} + \Delta$ where $\mathbf{I}$ is the identity. The collision operator differs from model to model and there are no restrictions on its definition at this stage.

The streaming operator $\mathcal{S}$ is a global operator that maps s on another configuration in X. It is defined by

$$(\mathcal{S}s)_i(\mathbf{x} + \mathbf{c}_i) = s_i(\mathbf{x}) \qquad (14.4)$$

where the $(\mathbf{c}_i)$ are any fixed set of vectors that connect sites on the Bravais lattice $\mathcal{L}$. (The notation $(\mathcal{S}s)_i$ in the above equation arises because $\mathcal{S}$ is a global operator that acts on s and sends it to $r = \mathcal{S}s$. This is to be contrasted with equation (14.3) where C is a local operator acting on $s(\mathbf{x}, t)$.) Thus this step of the dynamics is

$$s_i(\mathbf{x} + \mathbf{c}_i, t + 1) = s'_i(\mathbf{x}, t). \qquad (14.5)$$

We will assume in what follows that $\mathcal{L}$ is finite. To fix ideas, this will be realized by periodic boundary conditions, so that streaming lets particles leaving through one boundary reenter through the opposite boundary as in Figure 14.3. Altogether, noting $\mathcal{C}$ the global collision operator, the dynamics are

$$s(t+1) = \mathcal{S}\mathcal{C}s(t). \tag{14.6}$$

In the above equation, $\mathcal{S}$ and $\mathcal{C}$ are global operators acting on the configurations $s \in X$.

14.2 Probabilistic lattice gases

In the general case, the collision operator is not deterministic. The streaming step of the dynamics is unchanged. Given the local pre-collision configuration $\mathbf{n} = s(\mathbf{x})$ the post-collision $\mathbf{n}'$ is selected randomly among all possible configurations with a probability or *local transition rate* $A(\mathbf{n}, \mathbf{n}')$. The local transition rates thus define the collision operator, and they are tunable parameters of the models, although they should obey some restrictions. The transition rates should always sum to one:

$$\sum_{\mathbf{n}'} A(\mathbf{n}, \mathbf{n}') = 1 \quad \text{for all configurations} \quad \mathbf{n}. \tag{14.7}$$

They also must be bounded

$$0 \leq A(\mathbf{n}, \mathbf{n}') \leq 1. \tag{14.8}$$

A model will be called statistically reversible if it obeys the *detailed balance* condition

$$A(\mathbf{n}, \mathbf{n}') = A(\mathbf{n}', \mathbf{n}) \quad \text{for all} \quad \mathbf{n}, \mathbf{n}'. \tag{14.9}$$

However this condition is not easy to satisfy in practice in some 3D models, nor is it necessary to obtain most of the interesting effects of reversibility. A weaker condition is the condition of *semi-detailed balance* (SDB)

$$\sum_{\mathbf{n}} A(\mathbf{n}, \mathbf{n}') = 1 \quad \text{for all} \quad \mathbf{n}'. \tag{14.10}$$

This condition has many interesting consequences which we detail in the next section. In some developments, we request that the transition rates *preserve the symmetry of the lattice*. We let $\mathcal{G}$ be the group of point symmetries of the lattice as defined in Appendix A. Then if $\mathbf{R} \in \mathcal{G}$ is a symmetry of the lattice and we define $\mathbf{Rn}$ as the Boolean configuration $\mathbf{n}$ rotated by $\mathbf{R}$, transition rates obey

$$A(\mathbf{Rn}, \mathbf{Rn}') = A(\mathbf{n}, \mathbf{n}'). \tag{14.11}$$

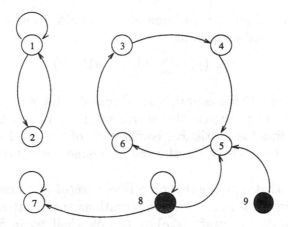

Fig. 14.2. A de Bruijn diagram for a Markov process. The arrows indicate non-zero transition probabilities. Notice that 8 and 9 are transient configurations. The ergodic components are the sets $\{7\}, \{1,2\}$ and $\{3,4,5,6\}$. One may easily imagine how period-4 probability distributions may arise over the set $\{3,4,5,6\}$.

14.3   Markov processes and the first H-theorem

Probabilistic lattice gases are an example of *Markov process*. The general framework for Markov processes is as follows. We still consider a finite set X of global configurations. There is a finite set of *transition rates* $a(s \to s')$ that equal the probability that the state s be transformed into s'. This defines a *Markov process* or a *probabilistic automaton*. Rates are probabilities and thus should obey

$$a(s' \to s) \geq 0 \quad \text{and} \quad \sum_s a(s' \to s) = 1. \qquad (14.12)$$

For a lattice gas, we define $a(s \to s')$ as the probability of s being transformed into s' by propagation followed by collision. Then $a(s \to s') = a_c(s \to S^{-1}s')$ where $a_c(r \to r')$ is the probability that r be transformed into r' by collisions. These *global collisional transition rates* a_c may be expressed simply as

$$a_c(s \to s') = \prod_{\mathbf{x} \in \mathcal{L}} A(s(\mathbf{x}), s'(\mathbf{x})). \qquad (14.13)$$

One may sketch de Bruijn diagrams for probabilistic cellular automata. One example is given in Figure 14.2. Unlike the deterministic case, states may have several outcomes.

   The full analysis of a Markov process aims at finding the probability $P_t(s)$ of observing configuration s at time t. The function $P_t(\cdot)$ is called the *state* of the lattice at time t in the classical terminology of statistical physics and should not be confused with a specific configuration s. The state obeys a Liouville equation

$$P_{t+1}(s) = \sum_{s'} a(s' \to s) P_t(s').$$ (14.14)

The full solution of this equation is of considerable complexity. In this chapter we will only discuss the *steady* solutions of the Liouville equation, which define the statistical equilibrium of the Markov process. To characterize the states, we first need to define some properties of the trajectories.

We will say that a configuration b is *connected* to a configuration c if there is a chain s^1, s^2, ...s^k of configurations such that $a(b \to s^1) > 0$, $a(s^1 \to s^2) > 0$, ..., $a(s^k \to c) > 0$. We shall write $b \to c$ when b is connected to c. A configuration b will be called a *transient* if there exists a configuration c such that $b \to c$ but not $c \to b$ (Fig. 14.2). If we now consider non-transient configurations, "being connected to" is an equivalence relation, and the equivalence classes are the *ergodic components*. In other words, two non-transient configurations are in the same ergodic component if and only if they are connected. Figure 14.2 has three ergodic components.

As a further example, a collision operator for the simple hexagonal lattice gas acts on the set X of the 26 possible configurations of a site. The de Bruijn diagram of this automaton is already too large to be conveniently represented, but one may notice that it splits in at least as many disjoint components as there are different mass-momentum packets. Figures 3.3 and 3.4 list all mass-momentum packets of the 7-velocity model. In some models, we split packet $(3,0,0)$ into two subpackets, which breaks into two separate ergodic components. On the other hand, for a random model, the mass-momentum packets are the ergodic components.

Probability is conserved so we may think of it as a fluid, and we sometimes call *mass* the probability of a configuration or a set. Over time, probability or mass flows from the transient states to the non-transient ergodic components. The long term solutions are thus restricted to the ergodic components. One may show that these long term solutions are either steady or periodic. To simplify the discussion in this chapter, we will only deal with steady solutions. For those an important theorem is the following:

Theorem 14.1 *For a single ergodic component there is a unique time independent solution of the Liouville equation.*

When there are multiple ergodic components, we may consider each ergodic component in isolation as a Markov system on its own. (No mass can flow out of the ergodic component, and in the long term there is no mass flowing *in* either since the transients have dried up.) Thus a steady or periodic probability distribution for a Markov process is unique once the mass of each ergodic component is fixed.

A particular case of interest occurs when the semi-detailed balance (SDB) property holds. SDB is inherited by the global rates when it holds for the local rates. From (14.13) and (14.10) we find

$$\sum_s a(s \to s') = 1. \tag{14.15}$$

An interesting fact is that when SDB holds there are no transient states. (The proof is left to the reader as Exercise 14.1.) We can then partition X into ergodic components and find the steady states. Once these are determined, we must investigate their stability. The problem of characterizing steady states is almost resolved, as witnessed by the following theorem:

Theorem 14.2 *Assume there is a single ergodic component, semi-detailed balance holds and all transition rates are positive. The unique long term solution is simply a constant probability over each ergodic component.*

The proof is in [14.1] and a full accounting of it is beyond the scope of this book. However, an important step in the proof is the famous *H-theorem*

Theorem 14.3 (H-theorem) *Let f be a convex function. Define H as*

$$H(t) = \sum_{s \in X} f[P_t(s)]. \tag{14.16}$$

If the system has SDB then H is a non-increasing function of time.

Proof: In equations, the theorem states that

$$H(t+1) \le H(t). \tag{14.17}$$

After using the Liouville equation (14.14) and conservation of probability (14.12) one may manipulate (14.17) to transform it into

$$\sum_{s'} \left\{ f\left[\sum_s a(s \to s') P_t(s) \right] - \sum_s a(s \to s') f[P_t(s)] \right\} \le 0. \tag{14.18}$$

For (14.18) to hold a sufficient condition is that each term in the curly brackets be non-positive. This may be written

$$f\left(\sum_s \lambda_s x_s \right) - \sum_s \lambda_s f(x_s) \le 0 \tag{14.19}$$

where $a(s \to s') = \lambda_s$ and $x_s = P_t(s)$. But we recognize that (14.19) is the definition of convexity whenever the λ_s are a set of nonnegative weights, i. e. $\lambda_s \geq 0$ and $\sum_s \lambda_s = 1$. This last equality holds because of semi-detailed balance (14.15). Thus (14.19) is verified and the theorem follows.   ∎

A particular case is obtained when one takes the function $f(x) = x \ln x$. It leads to the classical definition of the *entropy* of a state P as

$$H(P) = - \sum_s P(s) \ln[P(s)]. \tag{14.20}$$

14.4 Breaking of ergodicity

From the previous discussion it is clear that we would make a very significant progress if we knew the ergodic components. Ergodic components are closely related to *invariants* of the dynamics. An invariant is a function ϕ of $s \in X$ that stays constant in time, i. e.

$$\phi(s) = \phi(s') \quad \text{if} \quad a(s \to s') > 0. \tag{14.21}$$

Invariants are connected to ergodic components when there are no transient states. Thus we consider a system with no transients. (We may always consider the long time behavior when the transients are exhausted.) Then if $Z \subset X$ is an ergodic component of X, then the characteristic function of Z (i.e. $\phi(s) = 1$ if $s \in Z$ and $\phi(s) = 0$ otherwise) is an invariant. Conversely, if we know all the invariants, then we also know all the ergodic components.

The simplest situation would be to find mass and momentum to be the only invariants as they are intentionally built into the lattice gas. Indeed we may define a *global mass-momentum packet* $\Sigma_{M,\mathbf{G}} \subset X$ containing all configurations of mass M and momentum $\mathbf{G}$. This subset could be called an *energy surface* in analogy with the situation in classical statistical mechanics, or more accurately an *energy level* since the values of M and $\mathbf{G}$ are quantized in a finite configuration space X. Then we could say that *the system is ergodic if the ergodic components of X are the energy levels.* An ideal situation would be to have no unsteady periodic solutions of the Liouville equation but a single steady solution constant on each energy level. This could be expressed in the form of the following conjecture, which could be called an *ergodic conjecture for the lattice gas*:

Hypothesis 14.1 *At large times, the solutions of the Liouville equation converge to steady probability distributions which are constant over each energy level.*

Unfortunately one cannot prove this property and worse, it is easy to find counterexamples. Ergodic components that subdivide energy levels

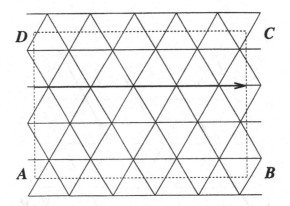

Fig. 14.3. A trivial non-ergodic state with a single particle on the hexagonal lattice. The system evolves on the domain ABCD with wrap-around boundary conditions. This simple trajectory does not cross all the configurations with the same momentum and mass. Hence it breaks ergodicity.

correspond to such violations of the ergodic conjecture. The associated probability distributions may be called non-ergodic states.

Typical non-ergodic states are obtained with a single particle as in Figures 14.3 and 14.4. We consider a $p \times 2q$ rectangular region of the infinite hexagonal lattice, with periodic boundary conditions. (The height must be a multiple of two for the wrap-around boundary condition to work naturally.) Since there is a single particle in the lattice there are no collisions and the momentum $\mathbf{G}_0$ of the particle stays constant. The energy level is $\Sigma_{1,\mathbf{G}_0}$. It is the set of all the configurations with the particle anywhere on the lattice with the same momentum. However the trajectory of the particle (which in this case is deterministic) does not cover the whole lattice. Instead it covers a smaller region of phase space, a typical non-ergodic situation. Figure 14.3 shows the situation for $\mathbf{G}_0 = (1,0)$, a particularly simple case. If we consider a particle with momentum $\mathbf{G}_0 = (\sqrt{3}/2, 0)$, there are several possible cases instead. When q divides p, the trajectory closes after winding around the domain a few times as in Figure 14.4. It is typically non-ergodic. On the other extreme, if p and q have no common divisor, the trajectory winds through all the nodes in the periodic domain and it is ergodic, as in Figure 14.5. Such examples of ergodicity breaking depend apparently on non-physical arithmetical details. However this is only an example of the much more complex violations of ergodicity that may happen with more particles.

Collisions may intuitively be expected to restore ergodicity, but they need not. For instance Figure 14.6 shows an example of a non-ergodic state in the square lattice model, involving collisions. There are clearly many more such non-ergodic states. How relevant are they on large lat-

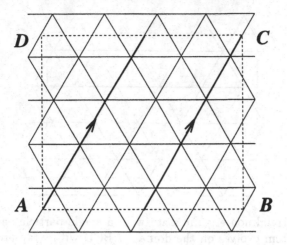

Fig. 14.4. Another non-ergodic state. The particle starts from A, winds twice around the domain and returns to A.

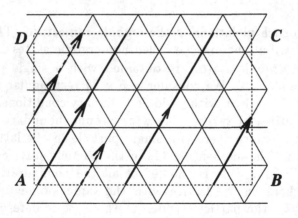

Fig. 14.5. An ergodic state on the hexagonal lattice. The particle winds around until it has crossed all the points in the domain.

tices with many particles? This is a difficult question that has not, to our knowledge, been fully addressed for the lattice gas. All that can be said is that many of the constructions of the classical statistical theory give correct results within the accuracy of numerical computations. Some of these constructions are quite elaborate so there is some reason to believe that the ergodic hypothesis is not too far off the mark, but again, this is only a conjecture.

Another step towards the answer is obtained if one considers only a linear kind of ergodicity. That is to say, we may restrict the search to invariants that are linear functions of the $s_i(\mathbf{x})$. Such *linear global invariants* will be studied in Section 15.7.

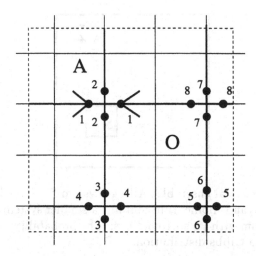

Fig. 14.6. An example of a non-ergodic state with two colliding particles. The particles start from A, with one of them rotating counterclockwise around O and returning to A, and the other making a similar turn (still counterclockwise) but crossing the boundaries of the periodic domain. Some configurations are never visited which proves non-ergodicity.

14.5 Gibbs states

The discussion of lattice-gas statistical mechanics, once the ergodic conjecture has been made, parallels exactly the discussion of classical or quantum statistics.

The statistical states or probability distributions that are steady solutions of the Liouville equation may be called statistical ensembles, following classical terminology. When we consider a system with a *fixed* energy level $\Sigma_{M,\mathbf{G}}$ we deal with the *microcanonical* ensemble. Let $|\Sigma_{M,\mathbf{G}}|$ be the number of elements in $\Sigma_{M,\mathbf{G}}$. Then the probability distribution is $P(s) = 1/|\Sigma_{M,\mathbf{G}}|$ for $s \in \Sigma_{M,\mathbf{G}}$ and $P(s) = 0$ otherwise. Besides the microcanonical ensembles, there are many steady states which are solutions of the Liouville equation. By the ergodic hypothesis, the steady states must be functions of the mass and momentum only, i. e.,

$$P(s) = f_P(M_s, \mathbf{G}_s), \qquad (14.22)$$

where M_s is the mass of configuration s and $\mathbf{G}_s$ its momentum. Among the various solutions, there is one that is most relevant physically and which corresponds to the limit of a *large* microcanonical ensemble Σ. Then a good approximation for a subset S of Σ is the *canonical ensemble* or *Gibbs distribution* (Figure 14.7). The energy of Σ is fixed, but we allow the energy of the system S to fluctuate as it exchanges energy with $\Sigma - S$,

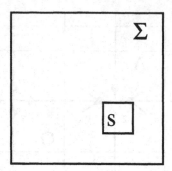

Fig. 14.7. The canonical ensemble. A large system Σ — the heat bath — with fixed values of mass and momentum contains a smaller system S that exchanges mass and momentum with its surroundings. The probability distribution for the smaller systems is a Gibbs distribution.

which is called the *heat bath*. (Although there is no heat in most of our models, we pursue the analogy between energy and lattice gas invariants.) The large lattice has a fixed energy, while any of the sublattices in Figure 14.7 is an instance of a canonical ensemble. The probability $P_S(s)$ to find a given sublattice S in a given state s may be computed exactly from the microcanonical probability distribution P_Σ for the large system Σ. In Appendix E we show how this probability distribution may be approximated for large heat baths Σ. The result is the *Gibbs distribution*

$$P_S(s) = \frac{1}{Z} \exp\left(-hM_s - \mathbf{q} \cdot \mathbf{G}_s\right), \tag{14.23}$$

where

$$Z = \sum_{s \in X_S} \exp\left(-hM_s - \mathbf{q} \cdot \mathbf{G}_s\right), \tag{14.24}$$

and h and $\mathbf{q}$ are adjustable parameters, called *chemical potentials*. In equation (14.23) all quantities are calculated for the canonical system S, and configurations belong to the configuration space X_S of S. An interesting property of the Gibbs distribution is that it is *factorized*. If A and B are two sublattices with configurations s_A and s_B, the joint configuration is $s = (s_A, s_B)$, and the joint probability distribution is $P(s) = P(s_A, s_B)$. If the joint probability distribution is a Gibbs distribution then it is the product of the Gibbs distributions for A and B:

$$P(s_A, s_B) = P_A(s_A)P_B(s_B). \tag{14.25}$$

The proof of (14.25) is easy, and simply involves the fact that the invariants are additive. For instance the total mass verifies

$$M_{(s_A, s_B)} = M_{s_A} + M_{s_B}, \tag{14.26}$$

along with a similar equation for $\mathbf{G}_s$. The additivity of the mass ensures the validity of equation (14.25) which simply means that the configurations of A and B are independent events. Interestingly, the converse is also true:

Property 14.1 *If an ergodic state leads to independent probability distributions for subsystems, it is a Gibbs state.*

To prove this fact, one sees from the ergodic hypothesis that equation (14.22) holds. But from independence we get the factorization property 14.1. Thus

$$f_P[M_{s_A} + M_{s_B}, \mathbf{G}_{s_A} + \mathbf{G}_{s_B}] = f_P[M_{s_A}, \mathbf{G}_{s_A}]f_P[M_{s_B}, \mathbf{G}_{s_B}]. \quad (14.27)$$

Since f_P transforms sums into products it must be the exponential function, up to a multiplicative constant Z^{-1}. This constant is determined by the property of summation of the probability ($\sum_s P(s) = 1$), yielding (14.24).

The Gibbs distribution is a powerful concept. It applies for any size of sublattice A, with the only restriction that the total number of particles $\mathcal{N}$ in the heat bath must be large. In usual macroscopic physics $\mathcal{N} \simeq 10^{23}$, of the order of the Avogadro number. However, for computational lattice gases the typical sizes are much smaller. It is thus interesting to estimate deviations from the Gibbs equilibrium. For this it is instructive to consider a very simple system,[*] made of $\mathcal{M}$ sites, with a Boolean variable $s(\mathbf{x}) \in \{0, 1\}$ on each site. The only invariant will be the total number of particles

$$\mathcal{N} = \sum_{\mathbf{x} \in \mathcal{L}} s(\mathbf{x}). \quad (14.28)$$

A schematic representation is given in Figure 14.8. We may imagine a number of probabilistic cellular automata for this system. Here we only require the ergodic hypothesis. To test the validity of the Gibbs distribution it is enough to consider two sublattices made each of a single site $A = \{\mathbf{x}_A\}, B = \{\mathbf{x}_B\}$. We shall write $s_A = s(\mathbf{x}_A), s_B = s(\mathbf{x}_B)$. The probability value is simply the expectation or statistical average $P_A(s_A = 1) = \langle s_A \rangle$. The factorization property then amounts to the disappearance of correlations:

$$\langle s_A s_B \rangle = \langle s_A \rangle \langle s_B \rangle. \quad (14.29)$$

---

[*] As an interesting terminological aside, such a system (in many cases with interactions added, see the notes section) is called a *lattice gas* in classical statistical physics. So the proper name for the models in this book should be momentum-and-mass-conserving lattice gases.

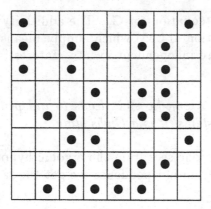

Fig. 14.8. A simple lattice-gas model, with only one particle per site.

Factorization is only approximately true. Indeed since all the $\mathcal{N}$ particles may be uniformly distributed on the $\mathcal{M}$ sites $\langle s_A \rangle = \mathcal{N}/\mathcal{M}$. But if we know there is already a particle on $\mathbf{x}_A$ there is one less particle to distribute on the rest of the lattice $\mathcal{L} - \mathbf{x}_A$. Thus the conditional probability distribution of finding a particle on B knowing there is a particle on A is

$$P(s_B = 1 | s_A = 1) = \frac{\mathcal{N} - 1}{\mathcal{M} - 1}. \qquad (14.30)$$

Since by definition of conditional probability

$$P(s_A, s_B) = P(s_B | s_A) P(s_A) \qquad (14.31)$$

we get

$$\langle s_A s_B \rangle = \frac{\mathcal{N} - 1}{\mathcal{M} - 1} \frac{\mathcal{N}}{\mathcal{M}}. \qquad (14.32)$$

The correlation coefficient may be defined as

$$C_{AB} = \frac{\langle s_A s_B \rangle}{\langle s_A \rangle \langle s_B \rangle} - 1. \qquad (14.33)$$

C_{AB} provides a measure of the correlations of sites A and B. When $C_{AB} = 0$ the sites are perfectly independent. Here we find

$$C_{AB} \simeq -1/\mathcal{N} + 1/\mathcal{M}. \qquad (14.34)$$

Thus factorization (and the Gibbs distribution) holds approximately if there is a large number of particles $\mathcal{N}$.

14.6 Fermi-Dirac distributions

The Gibbs distribution factorizes over the sites. This is a simple consequence of the factorization over subsystems (14.25). It may also be seen directly, as expression (14.23) is equivalent to

$$P(s) = \frac{1}{Z} \exp \left[- \sum_{\mathbf{x} \in \mathcal{L}} (hm_s(\mathbf{x}) + \mathbf{q} \cdot \mathbf{g}_s(\mathbf{x})) \right] \qquad (14.35)$$

where $m_s(\mathbf{x}) = \sum_i s_i(\mathbf{x})$ is the mass. Then

$$P(s) = \prod_{\mathbf{x} \in \mathcal{L}} P_{loc}[s(\mathbf{x})] \qquad (14.36)$$

where

$$P_{loc}(\mathbf{n}) = \frac{1}{Z_{loc}} \exp \left(-hm_{\mathbf{n}} - \mathbf{q} \cdot \mathbf{g}_{\mathbf{n}} \right). \qquad (14.37)$$

where $m_{\mathbf{n}} = \sum_i n_i$ is the mass of local configuration $\mathbf{n}$, with a similar notation for $\mathbf{g}_{\mathbf{n}}$. The constant Z_{loc} is a normalization constant which is determined by $\sum_{\mathbf{n}} P_{loc}(\mathbf{n}) = 1$ which yields

$$Z_{loc} = \sum_{\mathbf{m}} \exp \left(-hm_{\mathbf{m}} - \mathbf{q} \cdot \mathbf{g}_{\mathbf{m}} \right) \qquad (14.38)$$

where the sum is over local configurations. The distribution P_{loc} is a kind of "small" Gibbs distribution restricted to a canonical system S made of a single site. This distribution may be further factorized over the individual bits $s_i(\mathbf{x})$ such that

$$P_{loc}(\mathbf{n}) = \prod_i f_{FD,i}(n_i) \qquad (14.39)$$

where $f_{FD,i}$ is the *Fermi-Dirac* distribution:

$$f_{FD,i}(n_i) = \frac{\exp(-n_i h - n_i \mathbf{q} \cdot \mathbf{c}_i)}{1 + \exp(-h - \mathbf{q} \cdot \mathbf{c}_i)}. \qquad (14.40)$$

This yields the average populations $N_i^{eq} = f_{FD,i}(1)$:

$$
\begin{aligned}
N_i^{eq} &= \frac{\exp(-h - \mathbf{q} \cdot \mathbf{c}_i)}{1 + \exp(-h - \mathbf{q} \cdot \mathbf{c}_i)} \\
&= [1 + \exp(h + \mathbf{q} \cdot \mathbf{c}_i)]^{-1}. \qquad (14.41)
\end{aligned}
$$

We thus recover the Fermi-Dirac distribution (3.32) of Chapter 3.

14.7   A summary and critique

Our discussion of the foundations of statistical mechanics for lattice gases may be summarized by the following diagram:

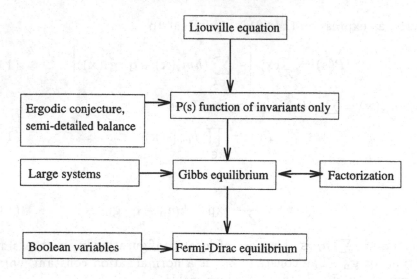

As seen in this diagram, the key step is the ergodic hypothesis. To investigate ergodicity, it is equivalent to search for the ergodic components or to search for all the invariants of the system. A brute force search is impractical: the number of configurations for a lattice gas grows at least like 2^{bM}, where b is the number of bits per site. This would rapidly overwhelm any computational facility. However we shall show in Section 15.7 that there are additional, linear, *staggered momentum invariants* for the hexagonal and FCHC models. Thus we may reformulate the ergodic hypothesis in the following way:

The only relevant invariants of the system are mass, momentum and the staggered invariants described in equation (15.81).

In other words the existence of non-ergodic states such as those described in Section 14.4 should lead to *no* observable deviations from the Gibbs distribution when we perform numerical experiments on equilibrium states. It is indeed likely that such non-ergodic states have little significance for the calculation of statistical properties. Another difficulty that may arise is that convergence to the ergodic, steady state distribution may be so slow that equilibrium is difficult or impossible to observe. This is what may happen if we attempted to observe spinodal decomposition in a "very viscous" model. Instead of observing the equilibrium, phase-separated state, we would see the mixed state almost "locked" in a non-equilibrium situation. (A simple way to observe this effect is for instance to simulate the ILG near saturation density or for otherwise very high viscosity.)

In general, statistical physics postulates a kind of gentle relaxation to equilibrium called linear response theory. This is what we describe in the following chapter.

14.8   Exercises

14.1 Show that there are no transients when SDB holds.

14.9   Notes

The discussion in this chapter is inspired to a large extent by:

14.1 Penrose, O. (1970). *Foundations of statistical mechanics. A deductive treatment* (Pergamon Press).

This chapter deals with equilibrium statistical properties of the lattice gas. Dynamical, or non-equilibrium, properties are discussed for instance in

14.2 Ernst, M. H. (1990). Linear response theory for cellular automaton fluids. In van Beijeren, H., editor, *Fundamental Problems in Statistical Mechanics*, VII, pages 321-355 (Elsevier).

14.3 Ernst, M. H. (1991). Mode-coupling theory and tails in CA fluids. *Physica D* **47**, 198–211.

15

Hydrodynamics in the
Boltzmann approximation

In this chapter we give a full derivation of the Navier-Stokes equation for the lattice gas. The first step, the Boltzmann approximation, is an approximation of the exact Liouville dynamics. The Boltzmann approximation should not be confused with the lattice-Boltzmann method of Chapter 6 and Chapter 7, but results that we obtain here for the Boltzmann approximation and the Navier-Stokes equations are also useful for the Boltzmann method. One of these results is the H-theorem for lattice gases. In this chapter we also reopen the tricky issue of the *spurious invariants* of lattice-gas or Boltzmann dynamics. We specifically discuss *non-uniform global linear invariants* for which, unlike the general nonlinear invariants, some theoretical results are known. Among them we find the *staggered momentum invariants*. We discuss their effect on hydrodynamics, which leads us to corrections to the Euler equation of Chapter 2.

15.1   General Boolean dynamics

It is useful to express the Boolean dynamics as a sequence of Boolean calculations, as we did in Section 2.6. In this section we shall denote by s or n local configurations. We further define a field of "rate bits", which are random, Boolean variables, defined independently on each site and denoted by $a_{ss'}(\mathbf{x}, t)$. They are equal to one with probability $\langle a_{ss'} \rangle = A(s, s')$. Thus if the pre-collision configuration is $\mathbf{n}$ the post-collision configuration is that single s' for which $a_{\mathbf{n}s'} = 1$. This remark leads to the following expression for post-collision configurations:

$$\mathbf{n} = \sum_{s'} a_{\mathbf{n}s'} s'. \tag{15.1}$$

Here new $a_{ss'}$ are drawn for each site $\mathbf{x}$ and time t. We rewrite this to read

$$n'_i = \sum_{s,s'} \delta_{\mathbf{n}s} a_{ss'} s'_i \tag{15.2}$$

where $\delta_{\mathbf{n}s}$ is a "Kronecker delta" that tests the equality of $\mathbf{n}$ and s,

$$\delta_{\mathbf{n}s} = \prod_j n_j(\mathbf{x},t)^{s_j} \overline{n}_j(\mathbf{x},t)^{\overline{s}_j}, \tag{15.3}$$

where we use again the notation $\overline{x} = 1 - x$. With further manipulation of this kind, and with the propagation step included, we get

$$n_i(\mathbf{x}+\mathbf{c}_i, t+1) = n_i(\mathbf{x},t) + \sum_{s,s'} a_{ss'}(\mathbf{x},t)(s'_i - s_i) \prod_j n_j(\mathbf{x},t)^{s_j} \overline{n}_j(\mathbf{x},t)^{\overline{s}_j}. \tag{15.4}$$

The right-hand side is a generalization of the products given in (2.35) and (2.37) with the Δ operator now

$$\Delta_i(\mathbf{x},t) = \sum_{s,s'} a_{ss'}(\mathbf{x},t)(s'_i - s_i) \prod_j n_j(\mathbf{x},t)^{s_j} \overline{n}_j(\mathbf{x},t)^{\overline{s}_j}. \tag{15.5}$$

For a given operator, it is interesting to investigate first the associated local invariants. The *local linear invariants* of a lattice gas are b-vectors $(a_i)_i$ such that

$$\sum_i a_i \mathbf{n}_i = \sum_i a_i \mathbf{n}'_i \quad \text{whenever} \quad A(\mathbf{n}, \mathbf{n}') > 0. \tag{15.6}$$

The conservation of mass corresponds to the b-vector $\mathbf{a} = (1, 1, \cdots, 1)$. Conservation of momentum corresponds to the b-vectors $(c_{1\alpha}, c_{2\alpha}, \cdots, c_{b\alpha})$ for $1 \leq \alpha \leq D$. We may imagine constructing models with additional physical invariants such as kinetic energy. This will be done in the following sections. Interestingly, some models may also have spurious linear collisional invariants. For instance model FHP-I with pair collisions alone conserves the invariant

$$\mathbf{a} = (1, -1, 1, -1, 1, -1). \tag{15.7}$$

Vectors $\mathbf{a}$ related to collision invariants form a vector space and we also loosely call them collision invariants. For a given model, one defines a basis set of linearly independent collision invariants. We will denote these basis vectors by $\mathbf{a}^{(k)}$. Once this basis is chosen, any invariant quantity may be expressed as a linear combination of collision invariants. Thus we have without further proof the following property:

Property 15.1 *Any vector a_i satisfying equation (15.6) may be written as a linear combination of collision invariants from the basis set*

$$\mathbf{a} = \sum_k \alpha_k \mathbf{a}^{(k)}. \tag{15.8}$$

It is relatively easy to enumerate all the local linear invariants for a given set of transition rates, and models are constructed with only a selected number of such invariants, for instance mass and momentum.

15.2   Boltzmann approximation and H-theorems

The important molecular chaos assumption of Boltzmann means that particles entering a collision are not correlated before they collide. In equations, for any combination of particles $a, b, \ldots, x$ entering a collision,

$$\langle n_a n_b \cdots n_x \rangle = \langle n_a \rangle \langle n_b \rangle \cdots \langle n_x \rangle. \tag{15.9}$$

Averaging equation (15.5) we find the post-collision populations to be

$$N_i' = N_i \Delta_i[\mathbf{N}] \tag{15.10}$$

where the Boltzmann collision operator Δ is defined by

$$\Delta_i[\mathbf{N}] = \sum_{s,s'} A(s, s')(s_i' - s_i) \prod_j N_j^{s_j} \overline{N}_j^{\overline{s}_j}. \tag{15.11}$$

Adding streaming, we find the *lattice-Boltzmann equation*

$$N_i(\mathbf{x} + \mathbf{c}_i, t + 1) = N_i(\mathbf{x}, t) + \Delta_i[\mathbf{N}(\mathbf{x}, t)]. \tag{15.12}$$

As before it is important to find the stationary solutions of this equation. We first consider homogeneous solutions $\mathbf{N}^{eq}$, which are independent of $\mathbf{x}$ and t. The homogeneous stationary solutions turn out to be the Fermi-Dirac distributions of Chapters 3 and 14. This is stated in the following theorem

Theorem 15.1 *Consider a lattice-gas model with semi-detailed balance. The following three statements are equivalent:*

1. The vector $\mathbf{N}$ *is a solution of the Boltzmann equation, i. e.*

$$\Delta_i[\mathbf{N}] = 0 \quad for\ all \quad i. \tag{15.13}$$

2. The following holds whenever $A(s, s') > 0$:

$$\prod_j N_j^{s_j} \overline{N}_j^{\overline{s}_j} = \prod_j N_j^{s_j'} \overline{N}_j^{\overline{s}_j'}. \tag{15.14}$$

3. The vector **N** *may be written as the Fermi-Dirac distribution*

$$N_i = \frac{1}{1 + \exp(h + \mathbf{q} \cdot \mathbf{c}_i)}.$$ (15.15)

The proof may be found in reference [15.1]. The three statements in the theorem give equivalent definitions of homogeneous equilibrium. Statements 1 and 3 are easily understood as in Section 3.4. Statement 2 means that all configurations in the same invariant packet have the same probability. A particular case is the identity (3.4).

Other important consequences of the Boltzmann approximation are the H-theorems. The first theorem is local and refers to the post-collision and pre-collision populations in equation (15.10):

Theorem 15.2 *In the Boltzmann approximation, and whenever semi-detailed balance holds,*

$$\sum_i N_i' \ln N_i' + \sum_i (1 - N_i') \ln(1 - N_i') \le \sum_i N_i \ln N_i + \sum_i (1 - N_i) \ln(1 - N_i).$$ (15.16)

This theorem leads to a global H-theorem. Define

$$H(t) = \sum_{\mathbf{x}} \left[\sum_i N_i(\mathbf{x}, t) \ln N_i(\mathbf{x}, t) + \sum_i [1 - N_i(\mathbf{x}, t)] \ln[1 - N_i(\mathbf{x}, t)] \right].$$ (15.17)

Then we have

Theorem 15.3 *In the Boltzmann approximation, and whenever semi-detailed balance holds the function $H(t)$ is non-increasing in time.*

The proof of these theorems, due to Hénon, is given in ref. [15.2].

15.3   Chapman-Enskog expansion

We now consider small deviations from equilibrium distributions as in Section 3.3. The densities ρ and $\rho\mathbf{u}$ fluctuate in space and time and so do the N_i, with the definitions

$$\rho(\mathbf{x}, t) = \sum_i N_i(\mathbf{x}, t)$$ (15.18)

$$\rho(\mathbf{x}, t)\mathbf{u}(\mathbf{x}, t) = \sum_i N_i(\mathbf{x}, t)\mathbf{c}_i.$$ (15.19)

In the first approximation, the populations are given by the Fermi-Dirac distribution, thus

$$N_i(\mathbf{x}, t) \simeq N_i^{(0)}(\mathbf{x}, t)$$ (15.20)

where

$$N_i^{(0)}(\mathbf{x}, t) = \frac{1}{1 + \exp\left[h(\mathbf{x}, t) + \mathbf{q}(\mathbf{x}, t) \cdot \mathbf{c}_i\right]} \tag{15.21}$$

and the chemical potentials h and $\mathbf{q}$ are found by inserting (15.20) and (15.21) into (15.18) and (15.19). We have already done this in equations (3.33) and (3.34). In the general case we get a low velocity expansion analogous to (3.52). Notice further that we use the notation $N_i^{(0)}$ instead of $N_i^{(eq)}$. Except for the fact that the former is varying in time and space and the latter is constant, there is no difference between the two.

The space and time scale of the fluctuations of ρ and $\rho\mathbf{u}$ is large, and thus derivatives —or gradients— of the $N_i^{(0)}$ are small. The Chapman-Enskog expansion is built on this idea. One writes

$$N_i = N_i^{(0)} + N_i^{(1)} + \cdots + N_i^{(n)} + \cdots \tag{15.22}$$

where the $N_i^{(n)}$ are small terms whose order of magnitude is described by the notation

$$N_i^{(k)} = \mathcal{O}(\nabla^k). \tag{15.23}$$

This means that the $N_i^{(k)}$ are of the same order as any space derivative of order k. It is assumed that the kth power of a space derivative is also of the same order, so the notation (15.23) may be interpreted to connect the $N_i^{(k)}$ to powers of space derivatives as well. Time derivatives are also small with $\partial_t = \mathcal{O}(\nabla)$. Moreover it will appear useful in what follows to couple the gradient expansion of equation (15.22) with a multiple time scale expansion. We let

$$\partial_t = \partial_{t_1} + \partial_{t_2} + \cdots \tag{15.24}$$

where $\partial_{t_1} = \mathcal{O}(\nabla)$, $\partial_{t_2} = \mathcal{O}(\nabla^2)$, etc.

Insertion of expansions (15.22) and (15.24) into the lattice-Boltzmann equation (4.6) produces at each order b equations. It is useful to introduce the linearized Boltzmann collision operator

$$\Lambda_{ij} = \left.\frac{\partial \Delta_i}{\partial N_j}\right|_{N=f\mathbf{1}}. \tag{15.25}$$

The above derivative is estimated at the zero-velocity equilibrium $\mathbf{N} \simeq f\mathbf{1} = (f, f, \cdots, f)$. To compute Λ_{ij} explicitly, notice first that

$$\frac{\partial}{\partial N_j}\left[N_j^{s_j}(1 - N_j)^{(1-s_j)}\right] = 2s_j - 1 \tag{15.26}$$

since s_j may be only 0 or 1. Then

$$\Lambda_{ij} = \sum_{ss'} (s'_i - s_i)(2s_j - 1)A(s, s') \prod_{k \neq j} N_k^{s_k} \overline{N}_k^{\overline{s}_k} \tag{15.27}$$

$$= \sum_{ss'} (s'_i - s_i)(2s_j - 1)A(s, s')f^{1-s_j}(1 - f)^{s_j} w(s) \tag{15.28}$$

$$= \sum_{ss'} (s'_i - s_i)(s_j - f)A(s, s')w(s) \tag{15.29}$$

where $w(s) = f^{n-1}(1 - f)^{b-n-1}$ and $n = \sum_i s_i$.

15.4 First order of the Chapman-Enskog expansion

At order 1 the following equation is obtained:

$$\partial_{t_1} N_i^{(0)} + c_{i\alpha}\partial_\alpha N_i^{(0)} = \sum_j \Lambda_{ij} N_j^{(1)}. \tag{15.30}$$

The operator Λ_{ij} is not invertible and a *solvability condition* needs to be satisfied in order to ensure the existence of a vector $\mathbf{N}^{(1)}$ obeying (15.30). As is the case in multiple-scale expansions in other branches of physics, the solvability conditions are associated with a continuous symmetry of the solutions or a conservation law, as we show below. From the conservation of mass and momentum, $D + 1$ vectors which are left-null-eigenvectors of Λ_{ij} are produced. These vectors are $\mathbf{1} = (1, ..., 1)$ and D vectors of the form $(c_{i\alpha})_{\alpha=1,\cdots,D}$. Indeed, mass conservation implies

$$\sum_i \Delta_i(\mathbf{N}) = 0 \tag{15.31}$$

for any $\mathbf{N}$. Hence setting $\mathbf{N} = \mathbf{N}^{eq} + \epsilon\mathbf{X}$ with $\mathbf{X}$ an arbitrary b-vector we find

$$\sum_{ij} \Lambda_{ij}X_j = 0 \tag{15.32}$$

for any $\mathbf{X}$ and thus $\mathbf{1}$ is a left-null-eigenvector of Λ_{ij}. Similarly momentum conservation implies

$$\sum_i c_{i\alpha}\Delta_i(\mathbf{N}) = 0 \tag{15.33}$$

and hence

$$\sum_{ij} c_{i\alpha}\Lambda_{ij}X_j = 0 \tag{15.34}$$

for any $\mathbf{X}$. Thus $c_{i\alpha}$ is a left-null-eigenvector of Λ_{ij}. We shall further assume that we have found the entire left-null space of Λ_{ij}. In other words, we assume that all the other eigenvalues of Λ_{ij} are non zero. This property is in general not too difficult to verify. For the 2D case we know all the eigenvalues from (4.43).

Multiplying (15.30) by 1 we get a consequence of the equations called a solvability condition:

$$\partial_{t_1} \sum_i N_i^{(0)} + \partial_\beta \sum_i c_{i\beta} N_i^{(0)} = 0. \tag{15.35}$$

Using the fact that the equilibrium distribution (15.21) has the required density one gets

$$\sum_i N_i^{(0)} = \rho \tag{15.36}$$

$$\sum_i N_i^{(0)} \mathbf{c}_i = \rho \mathbf{u}. \tag{15.37}$$

A further requirements on the Chapman-Enskog expansion is that no corrections to the mass and momentum appear at higher orders. Thus

$$\sum_i N_i^{(1)} = 0, \tag{15.38}$$

$$\sum_i N_i^{(1)} \mathbf{c}_i = 0. \tag{15.39}$$

From equations (15.35), (15.36), and (15.37) we obtain the mass-conservation equation

$$\partial_{t_1} \rho + \text{div} (\rho \mathbf{u}) = 0. \tag{15.40}$$

Multiplying (15.30) by the eigenvectors $c_{i\alpha}$ we obtain the momentum solvability condition

$$\partial_{t_1} \sum_i c_{i\alpha} N_i^{(0)} + \partial_\beta \sum_i N_i^{(0)} c_{i\alpha} c_{i\beta} = 0. \tag{15.41}$$

At this order the momentum flux is expressed as

$$\Pi_{\alpha\beta} = \sum_i N_i^{(0)} c_{i\alpha} c_{i\beta}. \tag{15.42}$$

The above expression is the momentum flux, or stress tensor, in local equilibrium. Expanding equation (15.41) and using equation (3.52), (3.53) we get the Euler equation

$$\partial_{t_1} \rho u_\alpha + \partial_\beta [g(\rho) \rho u_\alpha u_\beta] = -\partial_\alpha [p(\rho, u^2)] \tag{15.43}$$

where we have used the previous definitions of $p(\rho, u^2)$ and $g(\rho)$ given in equations (3.56) and (3.57). We thus recover our earlier result (3.55).

15.5   Second order mass conservation

First notice that the following identity holds:

$$\sum_i N_i(\mathbf{x} + \mathbf{c}_i, t + 1) = \sum_i N_i(\mathbf{x}, t). \tag{15.44}$$

These identities are simply the average of expressions (2.41) and (2.42). Since the populations N_i vary slowly in space and time, we may write the Taylor expansion

$$N_i(\mathbf{x} + \mathbf{c}_i, t + 1) = N_i + \partial_t N_i(\mathbf{x}, t) + c_{i\alpha}\partial_\alpha N_i(\mathbf{x}, t) + \mathcal{O}(\nabla^2). \tag{15.45}$$

Inserting into (15.44) and using (15.18) we get

$$\partial_t \rho + \partial_\alpha(\rho u_\alpha) + \mathcal{O}(\nabla^2) = 0 \tag{15.46}$$

which is only a rederivation of equation (2.49). However it is worth exploring the higher order $\mathcal{O}(\nabla^2)$ terms. The Taylor expansion is

$$N_i(\mathbf{x} + \mathbf{c}_i, t + 1) = N_i + \partial_t N_i + c_{i\alpha}\partial_\alpha N_i$$

$$+ \frac{1}{2}c_{i\alpha}c_{i\beta}\partial_\alpha\partial_\beta N_i + \partial_t c_{i\alpha}\partial_\alpha N_i + \frac{1}{2}\partial_{tt}^2 N_i + \mathcal{O}(\nabla^3) \tag{15.47}$$

where all terms on the right-hand side are evaluated at $\mathbf{x}, t$. Inserting into (15.44), and using (15.18), (15.19) and (15.42) we get

$$\partial_t \rho + \partial_\alpha(\rho u_\alpha) + \frac{1}{2}\partial_\alpha\partial_\beta \Pi_{\alpha\beta} + \partial_t\partial_\alpha(\rho u_\alpha) + \frac{1}{2}\partial_{tt}^2 \rho + \mathcal{O}(\nabla^3) = 0. \tag{15.48}$$

We now use the first-order conservation equations (15.41) and (15.46) to simplify the last three, second-order terms of the equations. Specifically from (2.49)

$$\frac{1}{2}\partial_{tt}^2 \rho = -\frac{1}{2}\partial_t\partial_\alpha(\rho u_\alpha) \tag{15.49}$$

and from (15.41)

$$\partial_t\partial_\alpha(\rho u_\alpha) = -\partial_\alpha\partial_\beta \Pi_{\alpha\beta}. \tag{15.50}$$

Using (15.49) and (15.50) equation (15.48) simplifies to

$$\partial_t \rho + \partial_\alpha(\rho u_\alpha) + \mathcal{O}(\nabla^3) = 0 \tag{15.51}$$

thus the mass conservation equation is *unchanged* at second order. This is quite remarkable, since it is not the case for momentum conservation.

15.6   Second order momentum conservation

Viscosity appears at the second order of the Chapman-Enskog equation. But before we write conservations equations at second order, we need to invert (15.30) for $\mathbf{N}^{(1)}$. To perform this inversion we need to be more specific about the operator Λ_{ij} defined in (15.25). We consider the case of a single rest particle, and we assume that the conditions for the symmetry of the linearized operator are fulfilled. The matrix Λ now has the form

$$\Lambda = \begin{pmatrix} \lambda_{01} & (\lambda_{1j}) \\ (\lambda_{1i}) & (A_{ij}) \end{pmatrix}. \tag{15.52}$$

The coefficients λ_{kj} and A_{ij} are simply a new notation for the coefficients Λ_{ij} given by equation (15.29). The coefficients A_{ij} involve moving particles only, λ_{01} involves the single rest particle, and the λ_{1j} describe the "coupling" between the single rest particle and moving particles. These coefficients obey two sets of constraints:

- *Angle dependence.* The matrix A_{ij} is invariant under the action of the symmetry group $\mathcal{G}$. In particular, the element A_{ij} depends only on the angle $(\mathbf{c}_i, \mathbf{c}_j)$. All coefficients λ_{1j} are equal.

- *Conservation laws.* Mass conservation expressed in equation (15.32) implies that

$$\sum_{j=1}^{b_m} A_{ij} + \lambda_{11} = 0 \tag{15.53}$$

and

$$\lambda_{01} + b_m \lambda_{11} = 0. \tag{15.54}$$

From momentum conservation as expressed in equation (15.34),

$$\sum_{j=1}^{b_m} \mathbf{c}_i A_{ij} = 0. \tag{15.55}$$

We first expand the left-hand side of equation (15.30). Using the equilibrium distribution (3.52) and (3.53), the first order mass-conservation law (15.40) and the Euler equation (3.47) itself, we find for moving particles

$$(\partial_t + c_{i\alpha}\partial_\alpha)N_{1i}^{(0)} = \left(\frac{D}{c^2 b_m}Q_{i\alpha\beta} + \frac{1}{bb_m}\delta_{\alpha\beta}\right)\partial_\beta(\rho u_\alpha) \tag{15.56}$$

and for rest particles

$$\partial_t N_{01}^{(0)} = -\frac{1}{b}\text{div}(\rho\mathbf{u}). \tag{15.57}$$

From the first-order equation (15.30), the above equations (15.56) and (15.57) and the definition (15.52) of the linearized operator,

$$\left(\frac{D}{c^2 b_m}Q_{i\alpha\beta} + \frac{1}{bb_m}\delta_{\alpha\beta}\right)\partial_\beta(\rho u_\alpha) = \sum_{j=1}^{b_m} A_{ij}N_{1j}^{(1)} + \lambda_{11}N_{01}^{(1)} \qquad (15.58)$$

$$-\frac{1}{b}\text{div}(\rho\mathbf{u}) = \lambda_{11}\sum_{j=1}^{b_m}N_{1j}^{(1)} - b_m\lambda_{11}N_{01}^{(1)}. \qquad (15.59)$$

We save a lot of effort if we first determine the general form of $\mathbf{N}^{(1)}$ from symmetry. Since the left-hand side of (15.59) depends on the symmetric part of $\partial_\beta(\rho u_\alpha)$, $\mathbf{N}^{(1)}$ must have the form

$$N_{1i}^{(1)} = t_{i\alpha\beta}^{(2)}\partial_\beta(\rho u_\alpha) + t_{i\alpha}^{(1)}\partial_\alpha\rho \qquad (15.60)$$

where $t_{i\alpha\beta}^{(2)}$ is a general tensor of second order, invariant by all lattice isometries in $\mathcal{G}$ that leave $\mathbf{c}_i$ invariant and symmetric in the Greek indices, and $t_{i\alpha}^{(1)}$ is similarly a vector invariant by all lattice isometries that leave $\mathbf{c}_i$ invariant. Also $N_{01}^{(1)} = K_{\alpha\beta}\partial_\beta(\rho u_\alpha)$ where $K_{\alpha\beta}$ is a general symmetric second order tensor invariant by all lattice isometries. In Appendix A we show that such a tensor must be of the form $K_{\alpha\beta} = Y\delta_{\alpha\beta}$ where Y is an arbitrary constant. We also show in Theorem A.1 that for all regular Bravais lattices

$$t_{i\alpha\beta}^{(2)} = \psi Q_{i\alpha\beta} - X\delta_{\alpha\beta}, \qquad (15.61)$$

and as before (Section 3.2) we have

$$t_{i\alpha}^{(1)} = Tc_{i\alpha} \qquad (15.62)$$

with arbitrary coefficients ψ, X, T. Using also the mass and momentum normalizations (15.38) and (15.39), we obtain $X = Y$ and $T = 0$. We then obtain

$$N_{1i}^{(1)} = (\psi Q_{i\alpha\beta} - X\delta_{\alpha\beta})\partial_\beta(\rho u_\alpha) \qquad (15.63)$$

$$N_{01}^{(1)} = b_m X\delta_{\alpha\beta}\partial_\beta(\rho u_\alpha) \qquad (15.64)$$

where the coefficients ψ and X depend on the operator Λ. Inserting these two equations into (15.59) and using (15.53) we find

$$X = \frac{1}{b^2 b_m \lambda_{11}}$$

$$\frac{D}{c^2 b_m}Q_{i\alpha\beta}\partial_\beta(\rho u_\alpha) = \psi\sum_{j=1}^{b_m}\Lambda_{ij}Q_{j\alpha\beta}\partial_\beta(\rho u_\alpha). \qquad (15.65)$$

This proves (albeit indirectly) that the b_m–vector $Q_{i\alpha\beta}$ is an eigenvector of A_{ij} (we recover the purely 2D result of Section 4.3). The corresponding eigenvalue is

$$\lambda = \frac{D}{c^2 b_m}\psi^{-1}. \tag{15.66}$$

Conversely, equations (15.63) and (15.64) yield the first order corrections in terms of λ.

We now turn to the second order of the Chapman-Enskog expansion. Substituting (15.22) into (4.6) we obtain at second order

$$\left[N_i^{(0)}(\mathbf{x}+\mathbf{c}_i,t+1) - N_i^{(0)}(\mathbf{x},t) + N_i^{(1)}(\mathbf{x}+\mathbf{c}_i,t+1) - N_i^{(1)}(\mathbf{x},t)\right]^{(2)}$$

$$= \sum_j \Lambda_{ij}N_j^{(2)} + \frac{1}{2}\sum_{jk}\frac{\partial^2\Delta_i}{\partial N_j\partial N_k}N_j^{(1)}N_k^{(1)} \tag{15.67}$$

where the notation $[\cdots]^{(2)}$ means that we collect all terms of order 2 in the expression in brackets. A solvability condition for momentum is obtained by multiplying equation (15.67) by the momentum eigenvector:

$$\sum_i \left[N_i^{(0)}(\mathbf{x}+\mathbf{c}_i,t+1) - N_i^{(0)}(\mathbf{x},t) + \right.$$

$$\left. N_i^{(1)}(\mathbf{x}+\mathbf{c}_i,t+1) - N_i^{(1)}(\mathbf{x},t)\right]^{(2)} c_{i\alpha} = 0. \tag{15.68}$$

We rewrite this expression going to second order in a Taylor expansion for the $N_i^{(0)}$ terms and to first order in a Taylor expansion for the $N_i^{(1)}$ terms:

$$\partial_{t_2}\sum_i c_{i\alpha}N_i^{(0)} + \frac{1}{2}\partial_{t_1}\sum_i c_{i\alpha}(\partial_{t_1}+c_{i\beta}\partial_\beta)N_i^{(0)}$$

$$+ \frac{1}{2}\partial_\beta\sum_i c_{i\alpha}c_{i\beta}(\partial_{t_1}+c_{i\gamma}\partial_\gamma)N_i^{(0)}$$

$$+ \partial_{t_1}\sum_i c_{i\alpha}N_i^{(1)} + \partial_\beta\sum_i c_{i\alpha}c_{i\beta}N_i^{(1)} = 0. \tag{15.69}$$

Simplifications occur using the two momentum normalization conditions (15.37) and (15.39) and we obtain

$$\partial_{t_2}\rho u_\alpha + \frac{1}{2}\partial_{t_1}\sum_i c_{i\alpha}(\partial_{t_1}+c_{i\beta}\partial_\beta)N_i^{(0)} + \frac{1}{2}\partial_\beta\sum_i c_{i\alpha}c_{i\beta}(\partial_{t_1}+c_{i\gamma}\partial_\gamma)N_i^{(0)}$$

$$+ \partial_\beta\sum_i c_{i\alpha}c_{i\beta}N_i^{(1)} = 0. \tag{15.70}$$

The second term is transformed using the first order in the Chapman-Enskog expansion, equation (15.30). The third term is transformed using the expansion of the streaming operator (15.56). We obtain

$$\partial_{t_2}\rho u_\alpha + \frac{1}{2}\partial_{t_1}\sum_{ij}\Lambda_{ij}c_{i\alpha}N_j^{(1)}$$

$$+ \frac{1}{2}\partial_\beta\sum_i c_{i\alpha}c_{i\beta}\left(\frac{D}{c^2 b_m}Q_{i\gamma\delta} + \frac{1}{bb_m}\delta_{\gamma\delta}\right)\partial_\gamma(\rho u_\delta)$$

$$+ \partial_\beta\sum_i c_{i\alpha}c_{i\beta}N_i^{(1)} = 0. \tag{15.71}$$

The second term disappears when we use (15.34). We replace the expressions (15.63) and (15.64) in the fourth term and use the fourth order tensor expression (3.46) again, to obtain

$$\partial_{t_2}\rho u_\alpha = \partial_\beta\left\{\nu[\partial_\beta(\rho u_\alpha) + \partial_\alpha(\rho u_\beta)]\right\} + \partial_\alpha\left[\nu_2\mathrm{div}(\rho\mathbf{u})\right] \tag{15.72}$$

where we introduce two viscosity coefficients

$$\nu = -\frac{b_m c^4}{D(D+2)}\psi - \frac{c^2}{2(D+2)} \tag{15.73}$$

$$\nu_2 = \frac{b_m c^2}{D}X + \frac{2b_m c^4}{D^2(D+2)}\psi + \frac{c^2}{D(D+2)} - \frac{c^2}{2bD}. \tag{15.74}$$

The coefficient ν is the usual kinematic shear viscosity. It can be reexpressed using equation (15.66) and (15.73)

$$\nu = -\frac{c^2}{\lambda(D+2)} - \frac{c^2}{2(D+2)}. \tag{15.75}$$

From equation (15.65), λ may be expressed as

$$\lambda = \frac{\sum_{ij}Q_{i\alpha\beta}A_{ij}Q_{j\alpha\beta}}{\sum_i Q_{i\alpha\beta}^2}. \tag{15.76}$$

The values of the shear viscosity for several classical models have been summarized in Table 4.1. The second coefficient ν_2 is related to the compression viscosity. Indeed by comparing equation (15.72) to the viscous terms in equation (2.30) we find that $\nu_2 = \xi/\rho$ when ρ is constant. When ρ is non-constant, it is not quite possible to identify the lattice gas equations with the fluid equations, because ρ is inside the derivatives instead of outside. This is another lattice effect that may be explained by the lack of Galilean invariance of the models.

Adding (15.72) to the Euler equation (15.43) and using the mutiple time scale expansion (15.24) we get the full Navier-Stokes equation of the lattice gas

$$\partial_t\rho u_\alpha + \partial_\beta[g(\rho)\rho u_\alpha u_\beta] = -\partial_\alpha[p(\rho, u^2)] + \partial_\beta\left\{\nu[\partial_\beta(\rho u_\alpha) + \partial_\alpha(\rho u_\beta)]\right\}$$
$$+ \partial_\alpha\left[\nu_2\mathrm{div}(\rho\mathbf{u})\right]. \tag{15.77}$$

Equations (15.77) and (15.51) are the main result of this chapter. Interestingly the form of these equations is universal, in the sense that it does not depend on the collision operator, except through the parameters ψ and X. These equations hold for all the lattice geometries whose symmetry properties imply isotropy of fourth order tensors. We have recognized several lattice effects. All of them disappear in the low Mach number limit. In that limit we can perform the same changes of variable as in Section 2.7 or Appendix D and obtain the incompressible Navier-Stokes equation.

The derivation we give above applies also to the lattice-Boltzmann method. There we can eliminate a number of lattice effects by an adequate choice of the pseudo-equilibrium populations, as shown in Section 6.2. However, some effects do remain, such as the fact that ρ is inside the derivatives in equation (15.72).

15.7   Non-uniform global linear invariants

The *global linear invariants* of a lattice gas are linear functions of the entire configuration s, of the form

$$L(s) = \sum_{\mathbf{x}} \sum_i a_i(\mathbf{x}) s_i(\mathbf{x}). \tag{15.78}$$

The field $a_i(\mathbf{x})$ characterizes the invariant L. Local invariants yield *uniform global invariants*, of the form

$$L(s) = \sum_{\mathbf{x}} \sum_i a_i s_i(\mathbf{x}) \tag{15.79}$$

where a_i is a local invariant defined in (15.6). To see that L is a global invariant, notice first that $\sum_i a_i s_i(\mathbf{x})$ is conserved by collisions. It is also conserved by streaming $\mathcal{S}$, since streaming only relocates the $s_i(\mathbf{x})$, leaving the sum over sites unchanged.

For a simple example of linear invariants distinct from the uniform global invariants, we turn to the square lattice gas of Section 3.1. Consider a horizontal line $\mathcal{L}_A$. Then the horizontal momentum on line $\mathcal{L}_A$ is

$$L_s = \sum_{\mathbf{x} \in \mathcal{L}_A} [s_1(\mathbf{x}, t) - s_3(\mathbf{x}, t)]. \tag{15.80}$$

L_s stays constant in time, because propagation keeps particles n_1 and n_3, and their momentum, on $\mathcal{L}_A$. Collisions send pairs of particles in and out of $\mathcal{L}_A$. Since these pairs carry a total momentum of 0, collisions do not affect L. The presence of line invariants explains why the viscous effects are very peculiar in the square lattice gas, as discussed in Section 4.1. In particular, momentum does not flow from one line to the next, and hence no momentum flux of the form Π_{xy} appears. For instance the flow (4.50) does not yield a correction to homogeneous equilibrium unlike the previous example (4.5).

Of particular interest are the staggered momentum invariants on the hexagonal lattice. Unlike the previous invariants, these involve a time dependence in their definition. They may be expressed as

$$J_s^{(k)} = \sum_{\mathbf{x}} (-1)^{t+\mathbf{b}_k \cdot \mathbf{x}} \mathbf{g}_s(\mathbf{x}) \cdot \mathbf{c}_k^{\perp} \tag{15.81}$$

for $k = 1$ to 3 where $\mathbf{c}_k^{\perp}$ is the unit vector orthogonal to $\mathbf{c}_k$, $\mathbf{b}_k$ is the reciprocal vector $\mathbf{b}_k = (2/\sqrt{3})\mathbf{c}_k^{\perp}$ and $\mathbf{g}_s(\mathbf{x}) = \sum_i s_i(\mathbf{x}, t)\mathbf{c}_i$ is the local momentum. Because of the time dependence, the $J_s^{(k)}$ are called *time-periodic invariants*. To verify that these expressions are indeed invariant, first notice that $\mathbf{g}(\mathbf{x}) \cdot \mathbf{c}_k^{\perp}$ is a projection of the momentum and thus invariant by the collision operator. The global expression in (15.81) is also invariant by the streaming operator S as easily seen by inspection*.

There are a series of results by Bernardin [15.4] that greatly simplify the search for invariants. The theory is based on the concept of regular models, which are defined as follows.

Definition 15.1 *A model with semi-detailed balance is called* regular *if every ergodic component is invariant by the streaming operator S.*

The motivation for the study of regular models comes from the fact that regularity immediately implies that the local coefficients $a_i(\mathbf{x})$ appearing in the expression (15.78) are local invariants of the collision operator. This arises because if a set is invariant by the whole dynamics SC and by S it must also be invariant by C alone. But since C acts locally on the configurations, the local coefficients $a_i(\mathbf{x})$ must be invariants of C.

It is easy to remark that all the linear invariants introduced so far are of the form (15.78), and the local coefficients $a_i(\mathbf{x})$ are always invariants of C. Thus regularity seems to hold for the cases considered so far. This is encouraging, since the search for invariants is much easier for a regular model. Indeed all we need to do is to determine those expressions of the form (15.78) which are invariant by streaming. However we did see an instance of non-regularity in a *non-linear* case: one may notice that the ergodic component defined in Figure 14.6 is not regular. It is thus of interest to determine rigorous mathematical conditions under which a model is regular. Bernardin shows that the following conditions are

---

* Strictly speaking, the invariants (15.81) are unchanged by streaming only when the dimensions of the domain are even integers in the direction $\mathbf{c}_k^{\perp}$. However, as we shall see below, the main effect of the invariants is to introduce a new slowly varying density, and this happens regardless of the size of the domain.

sufficient for a model to be regular:

$$\forall r, s \in X, a(r \to s) > 0 \Rightarrow a(s \to r) > 0. \qquad (15.82)$$

$$a(s \to r) > 0, a(q \to r) > 0 \quad \text{and} \quad q \neq s \Rightarrow a(s \to q) > 0. \qquad (15.83)$$

$$r \neq s \Rightarrow a(r \to s) < 1. \qquad (15.84)$$

Conditions (15.82) and (15.83) are satisfied by all the typical models.

Assuming regularity, the search for invariants may proceed in a systematic fashion. Invariants which have local coefficients a_i varying with short (2 or 3) space or time periods may be found by hand [15.4]. For larger periods a systematic search is still lacking, but other types of numerical searches using different ideas than regularity have been done [15.6, 15.5]. For the square lattice models, many linear invariants are found, including period 2 and period 3 invariants. For the hexagonal lattice models, however, no other linear invariants than mass, momentum and the three staggered momenta (15.81) have been found.

15.8   Equilibrium and hydrodynamics with additional invariants

It is straightforward to generalize expression (14.23) to account for the staggered momentum. One obtains

$$P_S(s) = \frac{1}{Z} \exp \left(-h M_s - \mathbf{q} \cdot \mathbf{G}_s - \sum_k \mu_k J_s^{(k)} \right), \qquad (15.85)$$

which is equivalent to

$$P(s) = \frac{1}{Z} \exp \left\{ - \sum_{\mathbf{x} \in \mathcal{L}} \left[h m_s(\mathbf{x}) + \mathbf{q} \cdot \mathbf{g}_s(\mathbf{x}) + \sum_{k=1}^{3} \mu_k j_s^{(k)}(\mathbf{x}, t) \right] \right\} \qquad (15.86)$$

where $m_s(\mathbf{x}) = \sum_i s_i(\mathbf{x})$ is the mass and $j_s^{(k)}(\mathbf{x}, t) = (-1)^{t + \mathbf{b}_k \cdot \mathbf{x}} \mathbf{g}_s(\mathbf{x}) \cdot \mathbf{c}_k^{\perp}$ is the local value of the kth staggered invariant. The constants h, $\mathbf{q}$ and μ_k are the chemical potentials for mass, momentum, and the $J^{(k)}$ invariants. Then the partition function is

$$Z = \sum_{\sigma \in X} \exp \left\{ - \sum_{\mathbf{x} \in \mathcal{L}} \left[h m_\sigma(\mathbf{x}) + \mathbf{q} \cdot \mathbf{g}_\sigma(\mathbf{x}) + \sum_{k=1}^{3} \mu_k j_\sigma^{(k)}(\mathbf{x}, t) \right] \right\}. \qquad (15.87)$$

The Gibbs distribution factorizes over the sites and the velocity indices as shown in Section 14.6. After some calculations, it may be shown that

$$P(s) = \prod_{\mathbf{x} \in \mathcal{L}} \prod_i f_{FD,i}(n_i(\mathbf{x})) \qquad (15.88)$$

where

$$f_{FD,i}(n_i) = \frac{\exp\left[-hn_i - \mathbf{q}\cdot n_i\mathbf{c}_i - \sum_{k=1}^{3}\mu_k(-1)^{t+\mathbf{b}_k\cdot\mathbf{x}}n_i\mathbf{c}_i\cdot\mathbf{c}_k^{\perp}\right]}{1 + \exp\left[-h - \mathbf{q}\cdot\mathbf{c}_i - \sum_{k=1}^{3}\mu_k(-1)^{t+\mathbf{b}_k\cdot\mathbf{x}}\mathbf{c}_i\cdot\mathbf{c}_k^{\perp}\right]}.$$
(15.89)

This yields the average populations $N_i^{eq} = f_{FD,i}(1)$:

$$N_i^{eq} = \left\{1 + \exp\left[h + \mathbf{q}\cdot\mathbf{c}_i + \sum_{k=1}^{3}\mu_k(-1)^{t+\mathbf{b}_k\cdot\mathbf{x}}\mathbf{c}_i\cdot\mathbf{c}_k^{\perp}\right]\right\}^{-1}.$$
(15.90)

Interestingly, if we let the chemical potential $\mu_k = 0$ we recover the previous Fermi-Dirac distribution (14.41).

The hydrodynamical equations in the presence of invariants involve the invariant densities defined by

$$\rho = \sum_i N_i$$
(15.91)

$$\rho\mathbf{u} = \sum_i N_i\mathbf{c}_i$$
(15.92)

$$h_k = (-1)^{t+\mathbf{b}_k\cdot\mathbf{x}}\sum_i N_i\mathbf{c}_i\cdot\mathbf{c}_k^{\perp}.$$
(15.93)

The third equation (15.93) is new and describes the density of staggered momentum $h_k(\mathbf{x},t) = \langle j^{(k)}(\mathbf{x},t)\rangle$. We also need to introduce the flux ϕ_k for the staggered momentum. We may not assume anymore that the population vary slowly in space and time, because the presence of staggered invariants implies precisely a rapid variation of the populations over a time and space period of two. However, it is still possible to proceed as in Chapter 4 to compute the average fluxes of mass, momentum and staggered invariants in terms of their densities. One starts by computing a low velocity, small h_k expansion of the N_i^{eq} at second order in a manner similar to (3.36). Then one may use the analogue of equation (15.44) (shown below) for the h_k to obtain a system of seven conservation laws

$$\partial_t\rho + \partial_\alpha(\rho u_\alpha) = 0$$
(15.94)
$$\partial_t(\rho u_\alpha) + \partial_\beta\Pi_{\alpha\beta} = 0 \quad \text{for} \quad 1 \le \alpha \le 2$$
(15.95)
$$\partial_t h_k + \partial_\alpha\phi_{k\alpha} = 0 \quad \text{for} \quad 1 \le k \le 3.$$
(15.96)

The previous expressions for the average mass and momentum flux are unchanged. As before, the momentum $\rho\mathbf{u}$ is also the mass flux and the momentum flux is still given by (2.51). The computation of the flux of staggered momentum ϕ_k is a bit more tricky. Notice that streaming and momentum conservation imply

$$\sum_i N_i(\mathbf{x},t)\mathbf{c}_i\cdot\mathbf{c}_k^{\perp} = \sum_i N_i(\mathbf{x}+\mathbf{c}_i,t+1)\mathbf{c}_i\cdot\mathbf{c}_k^{\perp}.$$
(15.97)

This expression is simply a statement of conservation of the staggered invariant by streaming. It is analogous to equations (15.44). Replacing the low velocity expansion of N_i^{eq} yields equation (15.96) with

$$\phi_k = \sum_i N_i(\mathbf{c}_k^\perp \cdot \mathbf{c}_i)\, \mathbf{c}_i \tag{15.98}$$

or

$$\phi_{k\alpha} = \Pi_{\alpha\beta} c_{k\beta}^\perp, \tag{15.99}$$

i.e., the component of momentum flux perpendicular to $\mathbf{c}_k$. Replacing in the conservation laws yields for a 7-velocity model on the hexagonal lattice

$$\partial_t \rho = -\partial_\alpha(\rho u_\alpha) \tag{15.100}$$

$$\partial_t(\rho u_\alpha) = -\partial_\beta \left[p\delta_{\alpha\beta} + \frac{g(\rho)}{\rho}\left(\rho^2 u_\alpha u_\beta + \sum_{k=1}^{3} c_{k\alpha}^\perp c_{k\beta}^\perp h_k^2 \right) \right] \tag{15.101}$$

$$\partial_t h_k = -\partial_\alpha \left[\frac{g(\rho)}{2\rho}\left(\delta_{\alpha\beta} + 2 c_{k\alpha}^\perp c_{k\beta}^\perp \right) \rho u_\beta h_k \right] \tag{15.102}$$

where

$$p = c_s^2 \rho - \frac{5}{14}\frac{g(\rho)}{\rho}\left(\rho^2 u^2 + \sum_k h_k^2 \right) \tag{15.103}$$

with $c_s^2 = 3/7$ and

$$g(\rho) = \frac{7}{12}\frac{1 - 2\rho/7}{1 - \rho/7}. \tag{15.104}$$

The equation for momentum is identical to the Euler equations (15.43) when the invariant densities are zero. However, when the invariant densities are non-zero, the momentum flux is modified through the terms involving h_k on the right-hand side of equation (15.101) and through the dependence of the pressure on h_k^2. In presence of non-zero staggered momentum, the hydrodynamics of the lattice are not equivalent to the Euler equations.

It is still possible to recover the Euler equations, however. In equation (15.102) for the staggered momentum densities h_k the fluxes are zero whenever the h_k are 0. Thus if the density of staggered momentum is initially zero, it stays zero and real-world hydrodynamics are safely recovered. Most methods for initializing a lattice gas simulation produce small invariant densities. But there are two caveats. Staggered momentum should not be produced at the boundaries. We have seen a simple example of boundary condition analysis in Chapter 7. If the boundary conditions suggested there are used, the invariant density at the boundary vanishes. A second caveat is to beware of shock waves. Indeed in a situation of rapid variation of mass and momentum, as in a shock wave, our assumption of slow variation of the invariants is no longer valid. There may thus be some generation of staggered momentum there also.

15.9 Exercises

15.1 Condition (15.84) is not true for the elementary models. To understand the meaning of condition (15.84) one may use as an example the situation in Figure 14.6. It depicts a closed deterministic trajectory.

1. Is this trajectory in itself an ergodic component? (Hint: the model is deterministic.)

2. By (careful) inspection it is possible to see that streaming does not leave the trajectory invariant. Explain how the trajectory is transformed when the streaming operator S alone (without collisions) is applied to configuration 1 in Figure 14.6.

3. Explain how the square lattice model could be changed to make condition (15.84) true.

4. What is the new ergodic component for the transformed model?

15.10 Notes

The proof of Theorem 15.1 may be found in:

15.1 Elton, B. H., Levermore, C. D. and Rodrigue, G. H. (1995). Convergence of convective-diffusive lattice Boltzmann methods. *SIAM J. Numer. Anal.* **32**, 1327–1354.

The proof of the H-theorems in this book may be found in:

15.2 Hénon, M. (1987). Appendix F. An H-theorem for lattice gases, *in:* Frisch, U., d'Humières, D., Hasslacher, B., Lallemand, P., Pomeau, Y., and Rivet, J.-P. (1987). Lattice gas hydrodynamics in two and three dimensions. *Complex Systems* **1**, 699-703.

The above article by Frisch et al. deals with the Chapman-Enskog expansion in the case of a fluid with no rest particles. It has inspired our account of the theory, but we chose to treat the single rest particle case. In more complicated models, where there are more rest particles or the set of allowed velocities contains more than nearest-neighbor lattice vectors (e.g. the 19 velocity model of Grosfils [3.5]) the expansion proceeds in a similar way. An explicit example may be found in [14.3].

The final expression we gave for the viscosity in equations (15.75) and (15.76) may be transformed in many ways, using semi-detailed balance and symmetry properties of Λ_{ij}. These transformations, and the proof that viscosity is positive, may be found in:

15.3 Hénon, M. (1987). Viscosity of a lattice gas. *Complex Systems* 1, 763–789.

The Bernardin approach to the search for invariants is from:

15.4 Bernardin, D. (1992). Global invariants and equilibrium states in lattice gases. *J. Stat. Phys.* **68**, 457–495.

Other attempts to search for invariants are described in:

15.5 d'Humières, D., Qian, Y., and Lallemand, P. (1990). Finding the linear invariants of lattice gases. In Pires A., Landau D. P., and Herrmann, H., editors, *Computational Physics and Cellular Automata*, pages 97–115 (World Scientific, Singapore).

15.6 Zanetti, G. (1991). Counting hydrodynamic modes in lattice gas automata models. *Physica D* **47**, 30–35.

The description of equilibrium and hydrodynamics in presence of stagger-ed-momentum invariants is inspired by:

15.7 Zanetti, G. (1989). Hydrodynamics of lattice gas automata. *Phys. Rev. A* **40**, 1539–1548.

16

Phase separation

Phase separation—the spontaneous separation of two initially mixed fluids—is one of the most fascinating aspects of immiscible lattice-gas mixtures. As we have already seen in Chapters 9, 11, and 12, phase separation occurs as a result of a *phase transition* in lattice-gas models. Though the resulting phase-separation dynamics can be quite dramatic, the transition itself can be difficult to describe. In particular, non-equilibrium aspects of lattice-gas phase transitions remain to be fully addressed.

Using a combination of theoretical arguments and numerical simulation, we focus in this chapter on the characterization of phase separation in lattice gases. We begin with a brief review of a classical model of phase separation. Then, focusing on the specific case of the immiscible lattice-gas (ILG) models of Chapters 9 and 11, we detail some of the ways in which our discrete models have been shown thus far to qualitatively (and sometimes quantitatively) reproduce the non-equilibrium evolution of real phase separation.

16.1   Phase separation in the real world

Phase separation occurs when the mixed state of a mixture is unstable, so that its components spontaneously segregate into bulk phases composed primarily of one species or the other. If the instability results from a finite, localized perturbation of concentration in the mixture, it is known as *nucleation*. If instead the perturbation is infinitesimal in amplitude, not localized, and of sufficiently long wavelength, the instability is known as *spinodal decomposition*.

A real mixture is said to be stable if, for a particular temperature T, its free energy F is at a minimum. Figure 16.1 illustrates how the free energy depends on temperature and the *order parameter* $\phi = \theta - 1/2$, where θ is the volume fraction of one of the mixture's components. The three curves

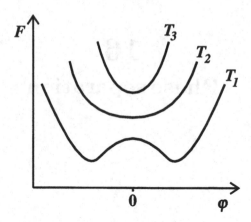

Fig. 16.1. Typical free-energy curves in a binary fluid mixture, for temperatures $T_1 < T_2 < T_3$, where $T_2 \approx T_c$.

correspond to temperatures $T > T_c$, $T \approx T_c$, and $T < T_c$, where T_c is the *critical temperature*. Above T_c, $F(\phi)$ has a single minimum, corresponding to a perfectly mixed state. In the simplest case of a symmetric binary fluid, this mixed state corresponds to a thermodynamic phase in which a single fluid consists of 50% of the "red" species and 50% of the "blue." Below T_c, however, the free energy curve develops a double well; the stable state now corresponds to a new thermodynamic phase in which a "red-rich" fluid coexists with a "blue-rich" fluid. The two fluids are separated from each other by interfaces, and the relative purity of each fluid (i.e., the redness of the red phase) increases with decreasing temperature.

The minima of the free energy curves of Figure 16.1 describe the equilibrium state of a mixture. We are particularly interested, however, in the non-equilbrium evolution from a mixed state to the unmixed state. A useful way to approach this problem theoretically is to consider the evolution at a spatial scale larger than that of the microscopic molecular motions but smaller than the hydrodynamic length scale L_{hydro} defined in Section 2.1. At this *mesoscopic* length scale the interfaces may no longer be described by jump conditions in a continuum-mechanical formulation (see Section 9.4) but instead have finite width. Thus we consider the order parameter $\phi(\mathbf{x}, t)$ to be slowly varying in both space and time. We then define the free-energy density $f_e(\phi)$, and understand that it, too, varies slowly in space and time.

For a symmetric binary fluid $f_e(\phi)$ is symmetric with respect to $\phi = 0$ and has the same form as $F(\phi)$ in Figure 16.1. The lowest order expansion of $f_e(\phi)$ that exhibits this qualitative behavior is

$$f_e(\phi) = -h_2\phi^2 + h_4\phi^4, \tag{16.1}$$

where $h_4 > 0$ and, for $T < T_c$, $h_2 > 0$. The total free energy F of the system is the integral over space of the sum of f_e and an additional spatial term chosen to favor smooth concentration fields. The resulting expression, known as a Ginzburg-Landau or Cahn-Hilliard free-energy functional, is

$$F[\phi] = \int d\mathbf{x} \left[\frac{\xi^2}{2} |\nabla \phi|^2 + f_e(\phi) \right],$$ (16.2)

where ξ is a parameter proportional to the width of interfaces. A dynamical model for the evolution of the order parameter is the continuity equation

$$\frac{\partial \phi(\mathbf{x}, t)}{\partial t} = -\nabla \cdot \mathbf{J}(\mathbf{x}, t),$$ (16.3)

coupled with an expression for the current $\mathbf{J}$ of the order parameter. This expression is obtained by noticing that the chemical potential μ_c is the variational derivative of F with respect to ϕ:

$$\mu_c(\mathbf{x}) = \frac{\delta F}{\delta \phi(\mathbf{x})}.$$ (16.4)

For small gradients, we expect the current $\mathbf{J}$ to be proportional to the gradient of the chemical potential. We thus obtain

$$\mathbf{J}(\mathbf{x}) = -M \nabla \mu_c$$ (16.5)

$$= -M \nabla \left(-\xi^2 \nabla^2 \phi + \frac{\partial f_e}{\partial \phi} \right),$$ (16.6)

where $M > 0$ is a mobility, or dissipation, coefficient. Substitution of equation (16.6) into equation (16.3) then yields the nonlinear evolution equation

$$\frac{\partial \phi(\mathbf{x})}{\partial t} = M \nabla^2 \left(-\xi^2 \nabla^2 \phi + \frac{\partial f_e}{\partial \phi} \right).$$ (16.7)

While equation (16.7) describes the evolution of the order parameter, we must analyze its stability with respect to perturbations from a uniform field ϕ_0 to determine the conditions necessary for phase separation to occur. We consider here only the case of spinodal decomposition. We write

$$\phi(\mathbf{x}) = \phi_0 + \tilde{\phi}(\mathbf{x}),$$ (16.8)

where $\tilde{\phi}(\mathbf{x})$ represents the perturbation from uniform concentration. Linearization of equation (16.7) about ϕ_0 then yields

$$\frac{\partial \tilde{\phi}(\mathbf{x})}{\partial t} = M \nabla^2 \left[-\xi^2 \nabla^2 + \left(\frac{\partial^2 f_e}{\partial \phi^2} \right)_{\phi_0} \right] \tilde{\phi}(\mathbf{x}).$$ (16.9)

In the initial stages of spinodal decomposition, one expects $\tilde{\phi}$ to be everywhere small. Thus, for sufficiently long wavelengths, the first term in the brackets above can be neglected, and we obtain the diffusion equation

$$\frac{\partial \tilde{\phi}(\mathbf{x})}{\partial t} = D_0 \nabla^2 \tilde{\phi}(\mathbf{x}), \tag{16.10}$$

where the diffusion coefficient D_0 is given by

$$D_0 = M \left(\frac{\partial^2 f_e}{\partial \phi^2} \right)_{\phi_0}. \tag{16.11}$$

We have now arrived at the principal result of this section. From equation (16.1), we know that the term in brackets in equation (16.11) may be either positive or negative. Thus the diffusivity D_0 may itself be positive or negative, and we see that equation (16.10) may describe either "downhill" or "uphill" diffusion. The case of uphill diffusion corresponds to the early stages of spinodal decomposition, since infinitesimal fluctuations from the uniform concentration ϕ_0 will grow rather than decay. We thus find that spinodal decomposition will occur whenever ϕ_0 and the temperature T are such that $D_0 < 0$.

Lastly, we discuss two additional concepts. First, we note that the locus of points for which $\partial^2 f_e / \partial \phi^2 = 0$ defines the *spinodal curve* indicated in Figure 16.2. Outside the spinodal curve the mixed phase is stable, while inside it is not. Second, we point out that the free-energy functional (16.2) does not include the effects of hydrodynamics, i.e., the conservation of momentum. More complicated thermodynamic potentials would be needed in this case.

16.2 Phase transitions in immiscible lattice gases

What does the theory of the previous section have in common with lattice-gas mixtures? One might be inclined to consider this mesoscopic theory irrelevant, because, for our microscopically-irreversible lattice-gas mixtures, (i.e., models lacking semi-detailed balance and thus not satisfying equation (14.10)), there is no compelling argument for the existence of a thermodynamical potential analogous to F. Nevertheless, we shall see that the behavior of the immiscible lattice gas offers some striking parallels with the linear instability derived above.

Specifically, we now show, both theoretically and empirically, that the phase separation instability in ILG's depends on the population density f and the concentration θ. The analysis results in a phase diagram in the plane of density and concentration, in which the phase boundary demarcates the two-phase, or phase-separated state, from the one-phase,

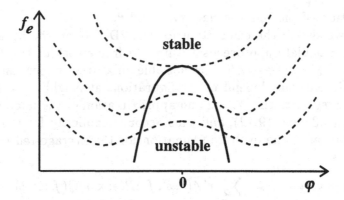

Fig. 16.2. Schematic diagram of a spinodal curve (smooth curve). The spinodal is defined by the locus of points for which $\partial^2 f_e / \partial \phi^2 = 0$, where the free-energy density f_e is indicated here by the dashed lines for three temperatures $T \geq T_c$. Outside the spinodal curve the mixed phase is stable, whereas inside it the mixed phase is unstable.

or mixed state. The phase diagram is analogous to the spinodal curve of Figure 16.2, where for the ILG the role of temperature is played by the density. As in the linear theory above, we show that the ILG spinodal curve may be obtained from the locus of points where the diffusivity $D_0(f, \theta) = 0$. Thus ILG phase separation results when f and θ are such that $D_0 < 0$.

To see why, we represent the current of concentration in the ILG not in terms of the variation of a potential F as in equation (16.6), but instead assume a Fickian, or linear, relation between concentration current and concentration gradient:

$$\mathbf{J} = -D_0(f, \theta) \rho \nabla \theta. \tag{16.12}$$

Here $\mathbf{J}$ is related to a coarse-grained average of the color flux $\mathbf{q}$ defined by equation (9.5), and D_0, the diffusion coefficient, depends on both the particle density and concentration. In lattice-gas models satisfying microscopic time-reversibility (or, more generally, semi-detailed balance), the H-theorem proved in Section 14.3 exists to show that transport coefficients are necessarily positive. The ILG collision rules, however, are time-irreversible and do not satisfy semi-detailed balance; thus D_0 may be either positive or negative. That D_0 can be negative is readily apparent from the rules expressed by equations (9.4), (9.5), and (9.6): as long as there are two colors present at a site, the collision rule *always* chooses to optimize the alignment of the flux $\mathbf{q}$ with the color gradient $\mathbf{f}$. The question that remains is whether $\mathbf{J}$ and $\nabla \theta$ may indeed be parallel on average. Thus we seek an estimate of D_0, or, more specifically, its sign, as a function of the reduced density f and θ.

Below we sketch the calculation for the 2D ILG on the hexagonal lattice. The calculation proceeds within the framework of the Boltzmann approximation as was done for miscible mixtures in Section 8.3. Because ILG collisions depend on configurations at neighboring sites, however, we must write the Boltzmann approximation of the microdynamical equations (9.3) and (9.12), and thus account explicitly for the neighboring populations. The evolution equation for the average red populations $R_i = \langle r_i \rangle$ is

$$R_i(\mathbf{x} + \mathbf{c}_i, t + 1) = \sum_{s,s',f_*} r_i' A(s, s', f_*) P(s; \mathbf{x}, t) Q(f_*; \mathbf{x}, t), \qquad (16.13)$$

while for the average blue populations $B_i = \langle b_i \rangle$ we have

$$B_i(\mathbf{x} + \mathbf{c}_i, t + 1) = \sum_{s,s',f_*} b_i' A(s, s', f_*) P(s; \mathbf{x}, t) Q(f_*; \mathbf{x}, t). \qquad (16.14)$$

Here the sums are taken over all possible states s that may enter a collision, all possible states s' that may result from a collision, and all possible discrete angle codes f_* as defined by equation (9.11). The factor $A(s, s', f_*)$ represents the probability of obtaining state s' when state s enters a collision at a site with a neighborhood configuration indexed by f_*. The probability that state s actually enters the collision at time t at the site located at position $\mathbf{x}$ is given by

$$P(s; \mathbf{x}, t) = \prod_{i=0}^{6} R_i^{r_i} B_i^{b_i} (1 - N_i)^{1 - r_i - b_i}. \qquad (16.15)$$

The probability that the discrete field angle is f_* is

$$Q(f_*; \mathbf{x}) = \sum_{(\phi_i): T(\phi_1, \ldots, \phi_6) = f_*} \left(\prod_{i=1}^{6} W(\phi_i; \mathbf{x}) \right), \qquad (16.16)$$

where the relative color density ϕ_i was defined in equation (9.9). Here the sum is taken over all possible combinations of $(\phi_i)_{i=1,\ldots,6}$ that correspond to f_* (via the operator T, as in equation (9.11)), and the product is taken over the probabilities $W(\phi_i)$ of observing the relative color density ϕ_i at the ith neighbor. Specifically, $W(\phi_i)$ is given by the sum of the probabilities of all states that yield the color density ϕ_i:

$$W(\phi_i; \mathbf{x}) = \sum_{s: M(s) = \phi_i} P(s; \mathbf{x} + \mathbf{c}_i). \qquad (16.17)$$

With these definitions the calculation of $D_0(f, \theta)$ may then proceed as in Section 8.3. (An alternative approach is a straightforward adaptation of the method used to calculate the ILG's surface tension in Chapter

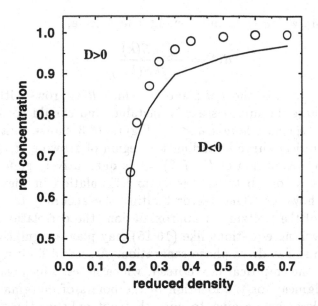

Fig. 16.3. Plot of Boltzmann approximation for $D(f,\theta) = 0$ (smooth curve) vs. empirical estimates of the point of marginal stability (circles) for ILG mixtures, in the plane of concentration θ and reduced density f [16.13]. Errors in the empirical estimates are approximately the same size as the symbols. The curves represent theoretical and empirical estimates of the ILG's analog of the spinodal curve of Figure 16.2.

17.) The calculation may not be performed analytically, however, because the ILG's collision tables (represented here by $A(s, s', S)$) are not easily represented by algebraic expressions. Thus the results are obtained by numerical solution of the resulting equations.

Figure 16.3 shows solutions to $D_0(f, \theta) = 0$. The region where $D_0 > 0$ corresponds to combinations of f and θ for which the mixed state is stable, while the region where $D_0 < 0$ corresponds to instability of the mixed state, and thus stability of the two-phase, or phase-separated, state.

Figure 16.3 also compares this theoretical estimate of the phase boundary to results of numerical *domain growth* experiments. In these experiments, the ILG is initialized as a homogeneous mixture for various combinations of f and θ, and the two-dimensional power-spectrum,

$$S(\mathbf{k}, t) = \left| \frac{1}{m} \sum_{\mathbf{x}} e^{-i\mathbf{k}\cdot\mathbf{x}} \phi(\mathbf{x}, t) \right|^2, \qquad (16.18)$$

is computed at discrete time intervals. Here $\phi = \theta - 1/2$ is the order parameter defined above, m is the number of lattice sites in the simulation, and $\mathbf{k}$ is the wave vector. To determine whether the mixed state is stable or unstable, the circular average $\hat{S}(k, t) = \langle S(\mathbf{k}, t) \rangle$, where $k = |\mathbf{k}|$, is computed, from which the time-varying length-scale,

$$R(t) = \frac{\sum_k \hat{S}(k)}{\sum_k k \hat{S}(k,t)} \tag{16.19}$$

is obtained. Regions in the f, θ plane for which $R(t)$ grows with time are the regions where the mixed state is unstable, and should correspond to the predicted regions where $D_0 < 0$. Figure 16.3 shows that comparison of the empirical curve bounding the region of growing $R(t)$ with the Boltzmann approximation of $D_0(f, \theta) = 0$ is qualitatively good.

Quantitative accord between theory and simulation in Figure 16.3 is not so good, however. One factor limiting the accuracy of the theory is the nature of the Boltzmann approximation: the correlations that are neglected in writing equations like (16.15) may play a significant role in the mechanisms that drive phase separation. A second limiting factor is the quality of the empirical curve itself, which necessarily involves some subjective judgment for the location of the points of marginal stability. It is nevertheless encouraging to note that not only do the two curves in Figure 16.3 give approximately the same critical density, $f_c \approx 0.2$ for $\theta = 1/2$, but so does the prediction of the surface tension coefficient discussed later in Chapter 17.

Thus we see that the early stages of spinodal decomposition in both the real world (i.e., equation (16.10)) and the ILG may be described as resulting from uphill diffusion. This raises a conundrum: real spinodal decomposition is driven energetically, while ILG spinodal decomposition is driven by its time-irreversible microdynamics. So may the two really be one and the same?

We have as yet no definitive answer. In the case of lattice gases it is probably more accurate to avoid thermal analogies and speak instead of bifurcations rather than phase transitions. Indeed, the language of *dissipative dynamical systems* is of some relevance here. We call "potential" those systems, like that of equation (16.7), that may be obtained from a thermodynamic potential, while we refer to all others as "non-potential." In the simplest case, a system $\phi(\mathbf{x}, t)$ derives from a potential F if it evolves such that

$$\partial_t \phi = -\frac{\delta F}{\delta \phi}. \tag{16.20}$$

We then have

$$\partial_t F = \partial_t \phi \frac{\delta F}{\delta \phi} = -(\partial_t \phi)^2, \tag{16.21}$$

or, expressed differently,

$$\partial_t F \leq 0. \tag{16.22}$$

Whenever F is a thermodynamic potential it has this non-increasing property—indeed it decreases in the presence of dissipative processes. The functional F may be both non-decreasing and non-thermodynamic, however. In this case it is called a *Lyapunov functional*. There is little numerical evidence that shows to which class—potential or non-potential—immiscible lattice-gas models belong. It would however be a remarkable accident if they fell into the potential class.

But accidents do indeed happen. Specifically, while we cannot guarantee that time-irreversible lattice-gas models will display the "correct" phase-separation dynamics, in the remainder of this chapter we catalog some empirical results concerning non-equilibrium pattern formation and growth where that is just what we find. Related questions concerning interface dynamics are left for Chapter 17.

16.3 Structure functions and self-similarity

One of the most common ways to study the time-dependent aspects of phase separation is to compute statistics from patterns such as those of Figures 9.2, 11.3, and 11.4. Probably the most comprehensive approach is to study the evolution of the power spectra $S(\mathbf{k}, t)$ given by equation (16.18). These spectra, which are sometimes called *structure functions*, have a characteristic shape and time dependence, some aspects of which may be predicted theoretically.

We consider below some examples of 3D ILG phase separation. A 2D slice of a 3D structure function has already been shown in Figure 11.5. Since the full 3D function $S(\mathbf{k})$ is approximately spherically symmetric, it is more convenient to average $S(\mathbf{k})$ over wavenumbers of equal length, i. e., for constant $k = |\mathbf{k}|$. In this way we obtain the spherical average $\hat{S}(k)$, analogous to the circular average defined in the previous section. The time-dependent characteristic size of domains, $R(t)$, is then once again given by the definition (16.19).

An example of the evolution of $\hat{S}(k, t)$ for a mixture with $\theta = 1/2$ is shown in Figure 16.4. As time increases, bubbles of each phase grow in size. Thus $R(t)$ grows, the power in the spectrum $\hat{S}(k, t)$ shifts to the left, and the distribution becomes more sharply peaked.

One expects that the dynamics of phase separation at any given time should depend on only one length scale—the characteristic domain size, $R(t)$. This is a statement of *self-similarity*: small bubbles should interact with small bubbles at early times in the same way that big bubbles interact with other big bubbles at late times. Thus the spectra of Figure 16.4 should follow the scaling law

$$\hat{S}(k, t) \propto F[kR(t)], \qquad (16.23)$$

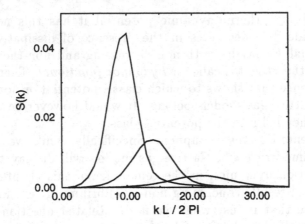

Fig. 16.4. Radially averaged power spectrum $\hat{S}(k,t)$ at time steps $t = 25$, 113, and 279 [16.14]. The average density is 20 particles per site and the average volume fraction of red is 50%. The 3D simulation is on a lattice of size $L = 128$. As time increases, the domains grow and the maximum of $\hat{S}(k)$ moves to the left.

where F is a scaling function and $R(t)$ is given by equation (16.19). The conservation of the order parameter ϕ allows more than a proportionality, since from Parseval's relation we expect that $\int S(\mathbf{k}, t)\mathrm{d}\mathbf{k} = \int \phi(\mathbf{x}, t)\mathrm{d}\mathbf{x}$ will be conserved. Spherical symmetry then implies that the left-hand side of (16.23) grows like R^2, while the rescaling of the argument of F gives another factor of R. We thus find that

$$\hat{S}(k,t) = R^3(t)F[kR(t)]. \qquad (16.24)$$

To test this prediction with the 3D ILG model, Figure 16.5 shows plots of $F[q(t)]$ versus $q(t)$, where the dimensionless wavenumber $q(t) = kR(t)$. We see that at late times $\hat{S}(k)$ is indeed self-similar as predicted by equation (16.24). At early times, however, we can see a slight drift of $F(q)$ with time. Thus it appears that somewhat different physical mechanisms operate in the early stages of ILG phase separation as compared to later times. Our assumption of self-similarity is therefore (unsurprisingly) naive, as one may equally expect for the case of real systems.

Figure 16.6 shows the same rescaled spectra, but now on a logarithmic plot so that some generic scaling properties may be better revealed. There are several features to note, all of which are known through some combination of theory, experiment, or other forms of numerical simulation. First, we see that for $q \gg 1$, $F(q)$ scales like q^{-4}. This scaling, well known from X-ray scattering studies, is known as *Porod's law*. It may be derived geometrically as a simple consequence of the existence of well-defined interfaces with thickness much smaller than $R(t)$. Next, as

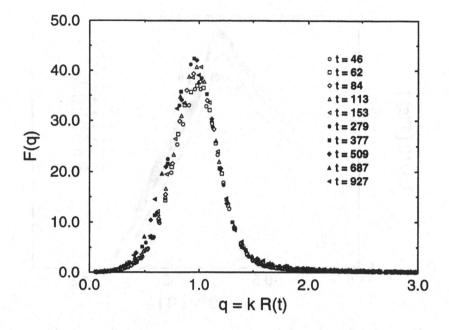

Fig. 16.5. Rescaled structure functions $F[q(t)] = R^{-3}(t)\hat{S}[q(t)]$ versus the dimensionless wavenumber $q(t) = kR(t)$ [16.14]. The simulations were performed with the 3D ILG model with $\theta = 1/2$ and $\rho = 20$ particles per site. At late times the data collapse nicely to one curve, validating the expected scaling law given by equation 16.24.

q decreases towards unity, the spectra reveal the effects of interfaces that are entangled at scales of order $R(t)$. Higher-order terms in an expansion of $F(q)$ then become important, and one expects scaling like $q^{-\alpha}$, with $\alpha \geq 6$ [16.9]. The crossover between this scaling and Porod's law typically occurs at $q \simeq 3$, as is also the case in Figure 16.6.

Another interesting crossover in the scaling of $F(q)$ occurs for $q < 1$. Figure 16.6 shows that $F(q) \sim q^2$ for $q < 0.3$, while $F(q) \sim q^4$ for $0.3 < q < 1$. These two scalings may be derived from a version of the evolution equation (16.7) that includes thermal noise [16.10]. Such an analysis shows that the q^2 scaling results from fluctuations at early times. As R becomes larger, however, the q^2 behavior is progressively less significant and it gives way to the q^4 behavior. That we see the crossover at $q \simeq 0.3$ is therefore one indication that the simulation has not yet reached the point where $R(t)$ is big enough for the effect of fluctuations to be unimportant.

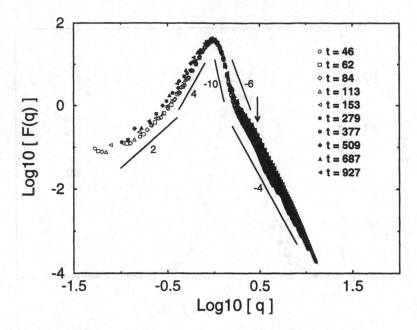

Fig. 16.6. Log-log plot of the same rescaled spectra shown in Figure 16.5 [16.14]. Numbers next to the straight lines give the slope of the line. For $q \gg 1$ we recover Porod's law q^{-4}. Other aspects of the scaling are discussed in the text. The arrow indicates the location $q = 3$.

16.4 Growth

It is also useful to study the evolution of the characteristic length scale $R(t)$ defined by equation (16.19). Figure 16.7 is a plot of $\log R$ versus $\log t$, made from simulations in boxes of size 64^3 and 128^3. Many comments may be made concerning this result.

First, one can see that the two curves are the same until about $t \simeq 1000$. Then, for the case of the small box, R grows no longer, while for the larger box it continues to increase. Though there is nothing surprising about this behavior, it nevertheless establishes that small boxes are sufficient to study the early-time growth.

Several physical mechanisms are responsible for the growth of $R(t)$. Of particular interest to us are the mechanisms of hydrodynamic origin. One simple and empirical way to establish the role of hydrodynamics is to break momentum conservation and to then compare growth with momentum conservation to growth without it, as shown in Figure 16.8. In the two non-hydrodynamic simulations, momentum conservation was broken by reversing particle velocities in a fixed percentage (10% and 20%) of collisions. We see that all three curves are the same for early

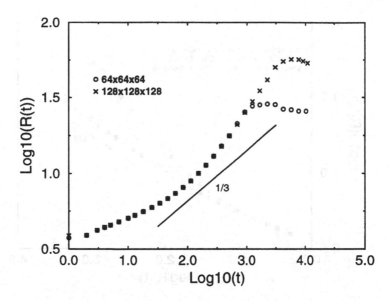

Fig. 16.7. Log-log plot of the characteristic size $R(t)$ made from simulations in boxes of size 64^3 and 128^3, for a volume fraction $\theta = 0.5$ and average density $\rho = 20$ [16.14]. Up until the time $t \simeq 1000$, both curves are the same, showing that finite-size effects are not significant in studying early-time growth. The straight line shown corresponds to $R \sim t^{1/3}$.

times, but that the hydrodynamic simulation quickly results in a faster growth rate.

The non-hydrodynamic growth in Figure 16.8 is compatible with the Lifshitz-Slyozov [16.6] growth law $R \sim t^{1/3}$ whereas the hydrodynamic growth in Figure 16.7 is not. Lifshitz-Slyozov growth derives from diffusion of one species through the other. It is unrelated to hydrodynamics and applies equally to systems such as binary alloys. The same exponent, however, may also be shown to result from coalescence due to the Brownian motion of droplets [16.7]. The latter mechanism is probably more relevant in ILG simulations.

Hydrodynamic effects are known to have a qualitative influence on growth when the volume fraction of the minority phase is large, as it is throughout this chapter ($\theta = 1/2$). When both phases are interconnected at late times, such as in the example of Figure 11.3, surface tension acts to reshape the contorted interfaces so that they become smoother and have less surface area. At the same time, viscous diffusion of momentum acts to slow down the growth due surface tension. Combining both surface tension and viscosity to form a length scale R_1, we find

$$R_1 \sim \frac{\sigma}{\mu} t, \tag{16.25}$$

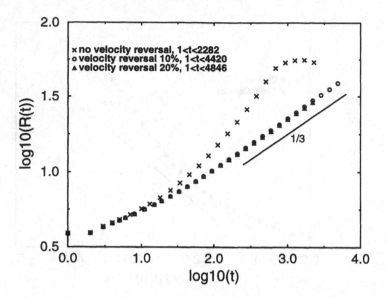

Fig. 16.8. Log-log plot of the characteristic size $R(t)$ with momentum conservation ($\times$'s) and without momentum conservation (circles and triangles) [16.14]. The latter cases correspond to reversing particle velocities in 10% and 20% of the collisions. Simulations were performed with an average density $\rho = 12$ and volume fraction $\theta = 0.5$. The cases without momentum conservation correspond well to the growth law $R \sim t^{1/3}$ at late times.

where $\mu = \rho\nu$ is the dynamic viscosity. We thus find the growth law $R_1 \sim t^1$, first predicted by Siggia [16.7].

Furukawa [16.8], on the other hand, has predicted another hydrodynamic regime in which inertial effects dominate viscous effects. Removing viscosity from the scaling (16.25), we find a length scale R_2 that grows like

$$R_2 \sim \left(\frac{\sigma}{\rho}\right)^{1/3} t^{2/3}. \tag{16.26}$$

There is unfortunately no clear indication in either Figures 16.7 or 16.8 that our momentum-conserving simulations have reproduced either of the two predicted hydrodynamic regimes. Both figures show late time growth that is roughly consistent with the $t^{2/3}$ law, but the choice of density in Figure 16.8 is such that the inertial term of equation (3.47) vanishes. Probably the size of our simulations is insufficient to see the effects of either growth mechanism. Nevertheless, it remains clear that hydrodynamics plays a strong role in the growth.

As we close this chapter, we wish to make two important points. First, we find it remarkable that the athermal ILG is able to reproduce as much of the physics of phase transitions as it does, even though this physics is usually considered to be of thermodynamic origin. Second, we wish to emphasize that the ability of lattice gases to simulate *both* hydrodynamics and phase separation is one of the most distinctive characteristics of these discrete models. Further studies of non-equilibrium growth in binary fluids will undoubtedly benefit from a more careful examination phase separation in lattice gases.

We now turn to the study of lattice-gas interfaces. We shall see there, as here, that our time-irreversible models simulate more physics than one would naively expect.

16.5   Notes

The theoretical model of phase separation in Section 16.1 is due to van der Waals and later to Cahn and Hilliard. A discussion of it may be found, for example, in:

16.1 Cahn, J. W. and Hilliard, J. E. (1958). Free energy of a nonuniform system. I. Interfacial free energy. *J. Chem. Phys.* **28**, 258–267.

16.2 Gunton, J., Miguel, M. S., and Sahni, P. (1983). The dynamics of first-order phase transitions. In Domb, C. and Lebowitz, J. L., editors, *Phase transitions and critical phenomena*, volume 8, pages 269–482 (Academic Press).

More complex models of phase separation and mixtures with several conserved quantities are discussed in:

16.3 Joseph, D. D. (1990). Fluid mechanics of two miscible liquids with diffusion and gradient stresses. *Euro. J. Mech. B: Fluids* **9**, 565–596.

16.4 Falk, F. (1992). Cahn-Hilliard theory and irreversible thermodynamics. *J. Non-Equilib. Thermodyn.* **17**, 53–65.

16.5 Nadiga, B. and Zaleski, S. (1996). Investigations of a two-phase fluid model. *Euro. J. Mech. B: Fluids* **15**, 885-896.

Dynamical aspects of phase separation are discussed in:

16.6 Lifshitz, E. M. and Slyozov, V. V. (1961). The kinetics of precipitation from supersaturated solid solutions. *J. Phys. Chem. Solids*, **19**, 35–50.

16.7 Siggia, E. D. (1979). Late stages of spinodal decomposition in binary mixtures. *Phys. Rev. A* **20**, 595–605.

16.8 Furukawa, H. (1985). Dynamic scaling assumption for phase separation. *Adv. Phys.* **34**, 703–750.

16.9 Tomita, H. (1986). Statistical properties of random interface systems. *Prog. Theor. Phys.* **75**, 482–495.

16.10 Yeung, C. (1988). Scaling and the small-wave-vector limit of the form factor in phase-ordering dynamics. *Phys. Rev. Lett.* **61**, 1135–1138.

The role of potential and non-potential models in dissipative dynamical systems is described in:

16.11 Pomeau, Y. (1986). Front motion, metastability, and subcritical bifurcations in hydrodynamics. *Physica D* **23**, 3–11.

16.12 Manneville, P. (1990). *Dissipative Structures and Weak Turbulence* (Academic Press, Boston).

The calculation of the diffusivity in the ILG is given in:

16.13 Rothman, D. H. and Zaleski, S. (1989). Spinodal decomposition in a lattice-gas automaton. *J. Physique France* **50**, 2161–2174.

The computations of structure functions and growth curves in the 3D ILG are from:

16.14 Appert, C., Olson, J. F., Rothman, D. H., and Zaleski, S.(1995). Phase separation in a three-dimensional two-phase hydrodynamic lattice gas. *J. Stat. Phys.* **81**, 181–197.

It is interesting to contrast the results of lattice-gas phase separation to those of the lattice-Boltzmann model of Chapter 10. Results from Boltzmann simulations are in:

16.15 Alexander, F. J., Chen, S., and Grunau, D. (1993). Hydrodynamic spinodal decomposition: growth kinetics and scaling functions. *Phys. Rev. B.* **48**, R634–R637.

Alexander et al. appear to quickly reach the hydrodynamic regime $R(t) \sim t^1$ in their 3D simulations. It may be that the fluctuations in the Boolean models are the cause of the qualitatively different growth in Figure 16.7. The role of fluctuations in momentum-conserving phase separation models remains to be adequately addressed, however.

The lack of semi-detailed balance is an issue that may be investigated in models that are simpler than the ILG. For example, the model may be simulated at density $f = 1$, thus making mass and momentum conservation trivial and moot. Such was the approach in:

16.16 Alexander, F. J., Edrei, I., Garrido, P. L., and Lebowitz, J. L. (1992). Phase transitions in a probabilistic cellular automaton: growth kinetics and critical properties. *J. Stat. Phys.* **68**, 497–514.

These authors recover the $t^{1/3}$ growth, perhaps for the same reason that it was found in the non-hydrodynamic simulations of Figure 16.8. They used probabilistic collision rules that simulated contact with a heat bath, similar to those introduced by:

16.17 Chan, C. and Liang, N. Y. (1990). Critical phenomena in an immiscible lattice gas cellular automaton. *Europhys. Lett.* **13**, 495–500.

Another way to investigate irreversible models is to consider their behavior in single-species gases. One example is the liquid-gas model of Chapter 12. Another, even simpler model, is in:

16.18 Bussemaker, H. and Ernst, M. H. (1993a). Lattice gas automata with self-organization. *Physica A* **194**, 258–270.

16.19 Bussemaker, H. and Ernst, M. H. (1993b). Pattern formation in a biased lattice-gas automaton. *Physics Letters A* **177**, 316–322.

These are just a few examples of models of *dynamical phase transitions*. They are principally distinguished by the conservation of momentum, and thus offer a richness not present in simple Ising systems. Their physics remains to be well understood.

17
Interfaces

Whereas phase separation is undoubtedly the most striking aspect of immiscible lattice-gas models, the interfaces that form due to phase separation are themselves an object of at least equal interest. In this chapter, we consider the interfaces formed by the 2D immiscible lattice-gas (ILG) model of Chapter 9.

We first discuss a theoretical calculation of the surface tension. Our calculation of surface tension not only provides a better understanding of ILG interfaces, but it also predicts the phase transition from the mixed to the unmixed phase described earlier in Chapter 16.

We then present a detailed view of interface fluctuations. In real fluids, interfaces fluctuate due to thermal noise. Lattice-gas interfaces, on the other hand, fluctuate due to the statistical noise in the Boolean dynamics. In both cases, the detailed motion of the interfaces results from a combination of surface tension, viscous hydrodynamics, and noisy excitation. We shall see that a study of interface fluctuations provides a delicate probe of the hydrodynamic and statistical properties of the ILG.

17.1 Surface tension: a Boltzmann approximation

The calculation of surface tension in ILG's offers neither the elegance nor the accuracy of the analogous calculation for lattice-Boltzmann models that we presented in Chapter 10. It does however yield several interesting results.

As in Section 10.2, we once again consider the surface tension of a flat interface. As we have already indicated in Exercise 11.3, there is no reason to expect that surface tension is isotropic in Boolean models. We do, however, expect that the anisotropy is six-fold symmetric as a result of the symmetry of the hexagonal lattice. As shown in Figure 17.1, there are two "generic" orientations of the interface, one parallel to a lattice line,

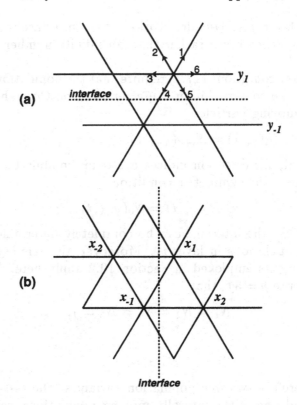

Fig. 17.1. (a) An interface parallel to a lattice line. (b) An interface perpendicular to a lattice line [1.19]. The directions $c_1, \ldots, c_6$ are labeled explicitly in the former case. Note that the parallel interface may be described by two layers y_1 and y_{-1} whereas that the perpendicular interface requires the four layers indicated by x_1, x_2, x_{-1}, and x_{-2}.

and the other perpendicular to a lattice line. The parallel case is simpler, because the minimum description of it requires only two parallel rows of the lattice, whereas the perpendicular case is staggered and requires four. Thus we shall describe in detail only the former case.

As shown in Figure 17.1, we take the center of the interface to be midway between and parallel to two horizontal lattice lines. Taking the y-axis to be vertical, we label the upper line y_1 and the lower line y_{-1}. Since the ILG model has one rest particle, there will be an average of $7f$ particles per site far from the interface. Thus we fix the boundary conditions

$$N_i(y_1, t) = f, \qquad i = 4, 5 \,\, \forall t \tag{17.1}$$

at all sites in layer y_1 and

$$N_i(y_{-1}, t) = f, \qquad i = 1, 2 \,\, \forall t \tag{17.2}$$

at all sites in layer y_{-1}. Our description of the interface is thus approximated by two lattice rows, rather than an infinite number as in Section 10.2.

The boundary conditions (17.1) require that the populations N_i be symmetric across the center of interface after rotation through 180 degrees. Thus for the moving particles

$$N_i(y_1, t) = N_{i+3}(y_{-1}, t), \qquad i = 1, \ldots, 6, \qquad (17.3)$$

where, as usual, the direction indices are read "modulo 6." For the rest particles we have the symmetric condition

$$N_0(y_1, t) = N_0(y_{-1}, t). \qquad (17.4)$$

Thus we see that the interface can be completely determined by solving only for the populations in layer y_1. Moreover, the same symmetry and stability arguments employed in Section 10.2 apply here. We then find that the pressure $p = 3f$, that

$$N_1 = N_2 = N_4 = N_5 = f, \qquad (17.5)$$

and that

$$N_3 = N_6. \qquad (17.6)$$

There are therefore two free population variables: the rest-particle population N_0 and one of the laterally moving populations, say N_3. As we do in our estimates of transport coefficients, we make the Boltzmann approximation that all populations are uncorrelated. Thus the populations evolve according to

$$N_i(y_1, t+1) = N_i'(y_1, t), \qquad i = 0, 3, 6, \qquad (17.7)$$

where the post-collision populations N_i' derive from the Boltzmann equation

$$N_i'(\mathbf{x}, t) = \sum_{s, s', f_*} n_i' A(s, s', f_*) P(s; \mathbf{x}, t) Q(f_*; \mathbf{x}, t), \qquad (17.8)$$

where the probabilities $P(s)$ of a state s and $Q(f_*)$ of an angle code f_* are given by equations (16.15) and (16.16), respectively, and the uncolored post-collision population $n_i' = r_i' + b_i'$.

Unlike the calculation in Section 10.2 we must now also specify the evolution of color, or concentration $\theta_i = R_i/N_i$. In addition to the symmetry given by equations (17.3) and (17.4), we also have

$$\theta_i(y_1, t) = 1 - \theta_{i+3}(y_{-1}, t), \qquad i = 1, \ldots, 6, \qquad (17.9)$$

and

$$\theta_0(y_1, t) = 1 - \theta_0(y_{-1}, t). \qquad (17.10)$$

Additional symmetries and stationarity of the interface allow the seven concentration variables in layer y_1 to be reduced to three independent variables. First, we note that symmetry under the reflection $x \rightarrow -x$ gives

$$\theta_3 = \theta_6, \qquad \theta_1 = \theta_2, \qquad \theta_4 = \theta_5. \tag{17.11}$$

Together with the concentration θ_0 for the rest-particle population, the first of these three pairs evolves according to

$$\theta_i(y_1, t+1) = \theta_i'(y_1, t), \qquad i = 0, 3, 6 \tag{17.12}$$

where we have used $\theta_i' = R_i'/N_i'$. The evolution of the second pair of concentrations is determined by particles that cross the interface; using equation (17.9), we obtain

$$\theta_i(y_1, t+1) = 1 - \theta_{i+3}'(y_1, t), \qquad i = 1, 2. \tag{17.13}$$

Since the stationarity of the interface requires that no net concentration crosses it, in steady state we must have

$$\theta_4'(y_1) = \theta_5'(y_1) = \frac{1}{2}, \tag{17.14}$$

and therefore, by equation (17.13),

$$\theta_1(y_1) = \theta_2(y_1) = \frac{1}{2}. \tag{17.15}$$

It remains only to specify the red concentration coming in from afar. Since in equilibrium the concentration that leaves the interface must be equal to the concentration that enters it, we set

$$\theta_i(y_1, t+1) = \theta_{i+3}'(y_1, t), \qquad i = 4, 5. \tag{17.16}$$

Thus two of the three free concentration variables may be taken to be θ_0 and θ_3, which enter via equation (17.12), while the third may be taken to be θ_4, which enters equation (17.16) above.

To complete the specification of the problem we need an expression for $Q(f_*)$, the probability of the angle code f_*. From equation (16.16), one sees that all that is required is knowledge of $W(\phi_i)$, the probability that the nearest-neighbor site in direction c_i has relative color density ϕ_i. These six quantities may each be obtained from the symmetries and boundary conditions of the problem. Taking $W(\phi_0)$ to be the probability distribution for relative color density for the interface site in layer y_1, one finds

$$W(\phi_3) = W(\phi_6) = W(\phi_0) \tag{17.17}$$

for the neighboring sites in layer y_1, and

$$W(\phi_4) = W(\phi_5) = W(-\phi_0) \tag{17.18}$$

for the neighboring sites across the interface in layer y_{-1}. For the sites on the boundary (i.e., layer y_2), one has, assuming that the chosen site in layer y_1 is at position $\mathbf{x}_1$,

$$N_i(\mathbf{x}_1 + \mathbf{c}_j, t) = f, \qquad i = 0, \ldots, 6, \qquad j = 1, 2 \tag{17.19}$$

for the populations, and

$$\theta_i(\mathbf{x}_1 + \mathbf{c}_j, t) = \theta_{j+3}(\mathbf{x}, t), \qquad i = 0, \ldots, 6, \qquad j = 1, 2 \tag{17.20}$$

for the concentrations. $W(\phi_1)$ and $W(\phi_2)$ may then be calculated directly from equations (16.15) and (16.17).

The evolution of the interface is thus fully specified by equation (17.7) for the two free populations N_0 and N_3, equations (17.12) and (17.16) for the three free concentrations θ_0, θ_3, and θ_4, and equations (17.17), (17.18), (17.19), and (17.20) for the determination of the probability of the angle code f_*. Due to the complexity of the ILG collision tables, this system of equations must be solved numerically by determining the steady-state post-collision populations N_i^*. The surface tension is then obtained from a symmetric two-layer version of equation (10.14):

$$\sigma = \sqrt{3} \sum_i (c_{iy}^2 - c_{ix}^2) N_i^* . \tag{17.21}$$

17.2   Boltzmann approximation versus simulation

Figure 17.2 compares results from the Boltzmann approximation for surface tension for both the parallel and perpendicular interfaces with results from three different empirical measurements made from simulations. We comment first on the theoretical predictions, and then on each of the empirical measurements.

Perhaps the most interesting feature of the theoretical calculation is the phase transition at $f = f_c \approx 0.2$. Below f_c surface tension vanishes, while above f_c the surface tension rises to a peak at about $f = 0.6$ and then falls to zero at $f = 1.0$. (Surface tension vanishes at $f = 1.0$ because each N_i must equal one.) One sees also that the perpendicular interface has a surface tension which is usually greater than that of the parallel interface, with a maximum deviation of about 20%.

The first of the three empirical measurements consists of the simulation of a red bubble of radius R in a blue box with periodic boundary conditions, of size greater than or equal to $4R$. The pressure p_1 inside the bubble is compared to the pressure p_2 outside the bubble. One expects adherence to Laplace's law (9.14) and a behavior with respect to radius

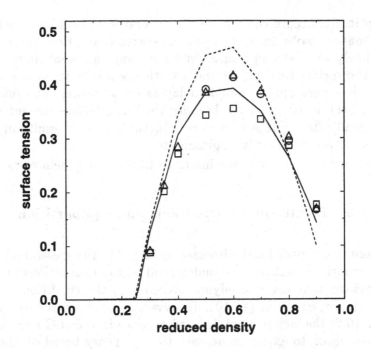

Fig. 17.2. Surface tension as a function of reduced density f in the 2D immiscible lattice gas [17.1]. Smooth curve: Boltzmann approximation for a two-layer interface parallel to a lattice line. Dotted curve: Boltzmann approximation for a four-layer interface perpendicular to a lattice line. Circles: empirical results from fitting Laplace's formula (9.14) to measurements made from bubbles of different sizes. Squares: measurements made from parallel interfaces that are flat on average. Triangles: measurements made from perpendicular interfaces that are flat on average. Error bars are smaller than the size of the symbols.

as shown previously in Figure 9.4. The results from a series of such measurements made at different densities ranging from $f = 0.5$ to $f = 0.9$ are indicated by the circles in Figure 17.2.

While the bubble tests should provide a measure of the average surface tension integrated over all angles, one may also make measurements of the surface tension from interfaces that are flat (on average) by numerical integration of equation (9.22). Figure 17.2 shows two such measurements, one set for a parallel interface, and the other for a perpendicular interface. These measurements display approximately the same magnitude of anisotropy that was determined from the Boltzmann approximation, with the perpendicular interface usually yielding the greater surface tension, as predicted.

Although the theoretical predictions and empirical measurements are always in qualitative agreement, quantitative agreement is lacking. The poor quantitative accord could be due to several reasons. First, the Boltzmann estimate was obtained by neglecting correlations. However, correlations are likely to be strong near an interface, and thus contribute significantly to the surface tension. Second, our theoretical estimate considered interfaces that were only two or four layers thick, rather than infinitely thick as in Section 10.2. Lastly, because the flat-interface measurements were not really flat, but were instead fluctuating, the results of these measurements are necessarily approximate.

We now turn to a study of the interface fluctuations themselves.

17.3   Equilibrium fluctuations and equipartition

As discussed in Chapter 14, the Boolean nature of lattice gases creates statistical fluctuations that may be understood as the fluctuations of Gibbs states provided that the microdynamics satisfies the condition of semi-detailed balance, equation (14.10). However, as we discussed at the end of Section 16.2, the lack of semi-detailed balance in the ILG implies that we have no reason to expect adherence to any theory based on the existence of a free energy. What kind of fluctuations, then, should we expect to see in the ILG? We consider below the case of interface fluctuations in the two-dimensional ILG.

It is useful to first review one of the predictions of classical statistical mechanics in systems that *do* satisfy detailed balance. In this case, interface fluctuations may be understood in terms of the fluctuations of surface energy. We consider one-dimensional interfaces $h(x,t)$ in a two-dimensional space, as shown in Figure 17.3. Since the energy E of a surface is proportional to its length L',

$$E = \sigma L' = \sigma \int_0^L dx \sqrt{1 + (\partial_x h)^2},   \tag{17.22}$$

where L is the length of the interface when it is flat and σ is once again the surface tension. When the gradient $\partial_x h \ll 1$, we can expand the argument of the square root in powers of the gradient and truncate at second order. We then obtain

$$E \approx \sigma L + \frac{\sigma}{2} \int_0^L dx (\partial_x h)^2.   \tag{17.23}$$

It is useful to decompose $h(x,t)$ into Fourier components of wavenumber $k = 2\pi n/L$, $n = 0, \pm 1, \pm 2 \ldots$, via the Fourier transform

$$h_k(t) = \frac{1}{L} \int_0^L e^{-ikx} h(x,t) dx   \tag{17.24}$$

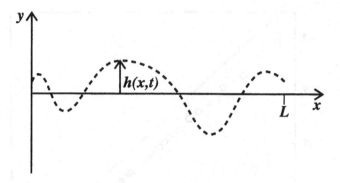

Fig. 17.3. Schematic diagram of the height $h(x, t)$ of an interface (dashed line) as a function of lateral position x and time t in a 2D fluid. When the interface is flat it has length L.

whose inverse is

$$h(x, t) = \sum_k e^{ikx} h_k(t). \tag{17.25}$$

Here the presence of the subscript k indicates that h is Fourier transformed, while otherwise it is not. Using these definitions, we find that equation (17.23) may be written

$$E = \sigma L + \frac{\sigma L}{2} \sum_k k^2 |h_k(t)|^2, \tag{17.26}$$

where we have substituted an equality for the approximation.

The *equipartition theorem* of statistical mechanics states that if the fluctuations of a system in equilibrium obey Gibbs statistics and the energy of that system may be decomposed into independent quadratic terms, then the average of each of the quadratic terms must be equal to $k_B T/2$, where T is temperature and k_B is Boltzmann's constant. For the case of equation (17.26), equipartition predicts the wavenumber power spectrum

$$\langle |h_k|^2 \rangle = \frac{k_B T}{\sigma L k^2}, \tag{17.27}$$

where the angle brackets imply an average over an ensemble of interfaces with different initial conditions, and the time dependence is dropped due to the assumption of equilibrium. We see then that the spectrum of spatial fluctuations decays like k^{-2}.

Figure 17.4 compares the prediction $\langle |h_k|^2 \rangle \sim k^{-2}$ with measurements made from simulations of ILG interface fluctuations. The ILG was initialized with a flat interface, with periodic boundary conditions in the direction parallel to the interface. After allowing time for the system to relax to equilibrium, the power spectrum $|h_k(t)|^2$ was computed from

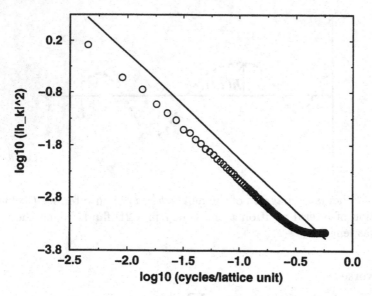

Fig. 17.4. Log-log plot of the power spectrum $\langle|h_k|^2\rangle$ obtained from simulations (circles) compared to a straight line with slope -2 [17.1]. Simulations were made with a 2D ILG using a lattice of size 256×256 and reduced density $f = 0.7$.

measurements $h(x, t)$ of the interface heights at each time step t, and then averaged over 10^5 time steps. Figure 17.4 shows $\log\langle|h_k|^2\rangle$ as a function of $\log k$, compared to a straight line with slope -2. Comparing the two curves, one finds that the slope of the empirical curve is indeed approximately -2 for wavenumbers below a high-wavenumber cutoff.

A more fundamental test of the equilibrium wavenumber spectrum of ILG interfaces would require an estimate of the amplitude of the ILG's noise; i.e., a quantity analogous to $k_B T$. It is unclear how to make such an estimate, however, due once again to the lack of a theoretical justification that the fluctuations are distributed in accord with a Gibbs distribution.

We are nevertheless still led to ask why the ILG has reproduced a prediction of equipartition when, due to its lack of semi-detailed balance, it appears to offer no basis for a such a theory. One possibility is that the interfaces are simply "rough" in the sense that the slopes $\partial_x h$ are random and uncorrelated. From this purely geometric assumption one readily finds from equations (17.24) and (17.25) that $|h_k|^2 \sim k^{-2}$.

However, one expects that a momentum-conserving model of interfaces contains more physics in its interface fluctuations than we have looked at so far. Thus in the next two sections we examine *time-dependent* aspects of the fluctuations. We shall see that not only are our expectations met, but that we will learn some interesting things about fluid interfaces at the same time.

17.4 Non-equilibrium roughening and dynamical scaling

A curious feature of interfaces in 2D fluctuating fluids is that they get rougher as they get longer. To see this, define the mean-square width $W^2(L)$ for an interface $h(x)$ with zero mean height:

$$W^2(L) = \frac{1}{L} \int_0^L h^2(x) \, dx. \qquad (17.28)$$

Using equation (17.27) and Parseval's relation $W^2(L) = \sum_k |h_k|^2$, we find

$$W^2(L) = \frac{k_B T}{\sigma L} \sum_{n=1}^{\infty} \left(\frac{2\pi n}{L}\right)^{-2} = \frac{k_B T}{12\sigma} L \qquad (17.29)$$

and therefore conclude that $W(L) \sim L^{1/2}$, i. e., that the root-mean-square width of an interface in equilibrium grows like the square-root of its length.

The power spectrum of Figure 17.4 has already established this scaling for the ILG. But we would now like to ask how W grows with time in addition to length.

We imagine a flat interface in a fluid that is preexcited by thermal noise. We assume that at time $t = 0$ the noise excites standing capillary waves of frequency ω_0, where ω_0 is given by the capillary-wave dispersion relation for inviscid fluids [17.8]:

$$\omega_0(k) = k^{3/2} \sqrt{\frac{\sigma}{2\rho}}. \qquad (17.30)$$

The wavenumber power spectrum should then grow like

$$\langle |h_k|^2 \rangle = \frac{2k_B T}{\sigma L k^2} \sin^2[\omega_0(k) t]. \qquad (17.31)$$

This prediction for the wavenumber spectrum incorporates both hydrodynamics and statistical mechanics. Note that by averaging equation (17.31) over time we also recover the equilibrium power spectrum (17.27).

To see how W grows with time, we once again apply Parseval's relation. From equations (17.30) and (17.31) we have

$$W^2(L, t) = \frac{2k_B T}{\sigma L} \sum_n \left(\frac{2\pi n}{L}\right)^{-2} \sin^2\left[\sqrt{\frac{\sigma}{2\rho}} \left(\frac{2\pi n}{L}\right)^{3/2} t\right], \qquad (17.32)$$

where we now explicitly note the dependence of W^2 on L and t. It is more revealing to write W^2 in the scaling form

$$W^2(L, t) = L \, g(t/L^{3/2}), \qquad (17.33)$$

where g is the scaling function

$$g(x) = \frac{k_B T}{2\sigma \pi^2} \sum_n n^{-2} \sin^2 \left[\sqrt{\frac{\sigma}{2\rho}} (2\pi n)^{3/2} x \right].$$ (17.34)

Equation (17.33) is a *dynamical scaling* relation. In conjunction with the scaling function g, it shows how the growth of the interface width depends on both time and the size of the system. At long times this inviscid form of $W(L, t)$ continues to oscillate since it does not provide for the viscous decay of capillary waves. In a viscous fluid, however, the initial excitations eventually decay, the phases of the excitation become random, and the interface on average reaches the equilibrium width predicted by equation (17.29). Thus the scaling function (17.34) is expected to be valid only for short times. Moreover, at short times we expect the dynamical scaling relation (17.33) to be independent of L. Hence $g(x)$ must have the form

$$g(x) \sim x^{2/3}$$ (17.35)

for small x, and $W^2(t) \sim t^{2/3}$ for small t.

To obtain a closed form expression for $W^2(t)$, we approximate the sum (17.32) by the integral

$$W^2(t) = \frac{k_B T}{\pi \sigma} \int_0^\infty dk\, k^{-2} \sin^2[\omega_0(k)t].$$ (17.36)

By making the substitution $x = \omega_0(k)t$, this integral takes the form

$$W^2(t) = \frac{2k_B T}{3\pi (2\rho\sigma^2)^{1/3}} t^{2/3} \int_0^\infty dx\, \frac{\sin^2 x}{x^{5/3}}.$$ (17.37)

The remaining integral may be evaluated analytically. One finds

$$\int_0^\infty dx\, \frac{\sin^2 x}{x^{5/3}} = \frac{\pi}{2^{1/3}\sqrt{3}\,\Gamma(5/3)} = 1.5947\ldots .$$ (17.38)

The early time growth of the interface is thus

$$W^2(t) = \frac{2^{1/3}}{3^{3/2}\,\Gamma(5/3)} \frac{k_B T}{(\rho\sigma^2)^{1/3}} t^{2/3}.$$ (17.39)

The width $W(t)$ is expected to grow like (17.39) until a crossover time t_c when the equilibrium state is attained. We may estimate t_c by assuming that $W(t)$ grows until the capillary wave with the longest wavelength has reached its maximum amplitude. This gives $t_c = T/4$, where the period $T = 2\pi/\omega_0$ is estimated from the capillary-wave dispersion relation (17.30) using $k = 2\pi/L$. We obtain the explicit estimate

$$t_c = \sqrt{L^3 \rho/(16\pi\sigma)}.$$ (17.40)

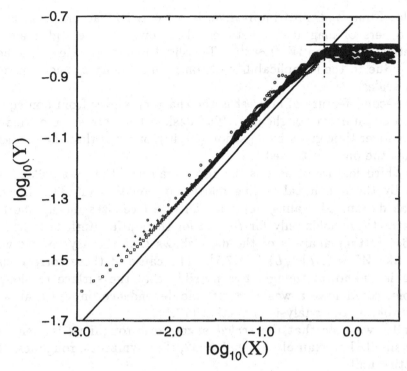

Fig. 17.5. Rescaled log-log plot of interface width W versus time t for interfaces of initial length L, where $Y = W/L^{1/2}$ and $X = t/L^{3/2}$ [17.7]. The straight smooth lines show the theoretical predictions given by equations (17.29) and (17.39). The dashed line shows the predicted crossover time t_c given by equation (17.40). The system sizes are $L = 16(\circ)$, $32(\circ)$, $64(\circ)$, $96(+)$, $128(\triangle)$, and $196(\square)$.

Figure 17.5 compares these predictions to measurements of $W(t)$ made from simulations of ILG interfaces. The simulations were initialized with flat interfaces of lengths ranging from $L = 16$ to $L = 192$. In the plot, both the width and time are rescaled in accordance with the prediction of dynamical scaling giving by equation (17.33). The predictions for the time dependent growth, equation (17.39), and the equilibrium width, equation (17.29), are given by the smooth straight lines. Both predictions use the same value of $k_B T$. This *effective temperature* (or, more formally, the effective energy per degree of freedom) is denoted by $k_B T_{\text{eff}}$. It is obtained empirically from independent simulations described in the next section. The theoretical predictions contain no other free parameters—the value of the surface tension was fixed by the choice of density ($\rho = 4.9$) and the prediction made from the Boltzmann approximation described in Section 17.1.

Four features of the roughening and consequent equilibrium state are notable. First, and most impressive, is the excellent accord, over almost three orders of magnitude, between the growth of the width and the theoretical prediction $W(t) \sim t^{1/3}$. The slight underestimate of the theory may be due to the inapplicability of continuum theory at short time and length scales.

The second feature of interest is the sharp crossover from non-equilibrium to equilibrium roughening. The dashed line shows the estimate of the crossover time given by equation (17.40); one sees that the prediction matches the data quite well.

The third feature of note is the good quality of the data collapse predicted by the dynamical scaling relation of equation (17.33). However, whereas dynamical scaling holds for all interface sizes during the transient growth, it holds only for $L \leq 64$ for the static interface width. For $L > 64$, further analysis of the data shows that the asymptotic width scales like $W \sim (L/\log L)^{1/2}$ [17.7]. The cause of these corrections to scaling is unknown, though it is possible that the surface tension, for example, could have a weak logarithmic dependence on size that is not predictable by our analysis in Section (17.1).

Finally, we note that the scaled asymptotic roughness is seen to be rather small. For example, when $L = 32$, the asymptotic roughness $W \approx 0.8$ lattice units.

17.5 Fluctuation-dissipation theorem and the frequency spectrum

A fuller understanding of time-dependent fluctuations that goes beyond the previous section is complicated by the coupling between thermal excitations in the fluid bulk and the motion of the interface. If there were no interface, the thermal noise in the bulk would give rise to fluid motions in the bulk that would be uncorrelated in space and time. In the presence of a moving interface, however, the hydrodynamic coupling between the motion in the bulk and the motion of the interface causes space and time correlations to arise. In other words, fluctuations in the fluid cause the interface to move which in turn causes the fluid to move, etc. These correlations may be described in terms of hydrodynamic response functions that give the response of the system to an external force.

The problem may be analyzed in terms of a *fluctuation-dissipation theorem*, which states that fluctuations of the interface, when averaged to a correlation function, decay in the same way as a macroscopic perturbation of the interface. Mathematically, this may be written as

$$\frac{\langle \dot{h}_k(t)\dot{h}_k^*(0)\rangle}{\langle |\dot{h}_k(0)|^2\rangle} = \frac{\dot{H}_k(|t|)}{\dot{H}_k(0)}, \tag{17.41}$$

where the interface velocities $\dot{h}_k(t)$ on the left-hand side are due only to thermal fluctuations in the fluid, while, on the right-hand side, the velocities $\dot{H}_k(t)$ are taken as non-fluctuating macroscopic (hydrodynamic) quantities. The fluctuation-dissipation theorem thus states that the problem of obtaining the correlation function of the height-velocity fluctuations, $\langle \dot{h}_k(t)\dot{h}_k^*(0)\rangle$, is, within a prefactor, a purely hydrodynamic problem.

To compare this theory to simulation, it is more convenient to compute the frequency power spectrum $|\hat{h}_k(\omega)|^2$, where $\hat{h}_k(\omega)$ is defined by the Fourier transform

$$\hat{h}_k(\omega) = \frac{1}{2\pi} \int_{-\Theta/2}^{\Theta/2} dt \, h_k(t)e^{-i\omega t}, \qquad (17.42)$$

where Θ is the size of the time-integration domain. The hydrodynamic problem on the right-hand side of the fluctuation-dissipation theorem (17.41) may be solved by deriving the solution to the linearized equation of motion for the interface. Such a solution takes the general form

$$\dot{H}_k(t) = \frac{1}{2\pi} \int_{-\infty}^{t} dt' \, R_k(t-t') \, F_k(t'), \qquad (17.43)$$

where $F_k(t)$ is the kth Fourier component of an arbitrary force acting on the fluid at the position of the interface, and $R_k(t)$ is a hydrodynamic response function. The frequency-domain expression for R_k is given below. The most important physical characteristic of R_k is that it is non-Markovian. In other words, the interface motion in the future depends not only on the present state of the interface, but also on the motion of the interface in the past. This is a consequence of the aforementioned coupling between the interface motion and the viscous flow in the bulk.

The force $F_k(t)$ is composed of surface tension effects and other forces F_k' such that

$$F_k(t) = -\sigma k^2 H_k(t) + F_k'(t). \qquad (17.44)$$

Substituting this expression into the equation of motion (17.43), we obtain

$$\dot{H}_k(t) = \frac{1}{2\pi} \int_{-\infty}^{t} dt' \, R_k(t-t') \, [-\sigma k^2 H_k(t') + F_k'(t')]. \qquad (17.45)$$

Using the Fourier transform relation (17.42), we may express this convolution integral as a product in the frequency domain:

$$\dot{\hat{H}}_k(\omega) = \hat{R}_k(\omega) \, [-\sigma k^2 \hat{H}_k(\omega) + \hat{F}_k'(\omega)]. \qquad (17.46)$$

The hydrodynamic response function $\hat{R}_k(\omega)$ is derived in Appendix F. It takes the form

$$\hat{R}_k(\omega) = \frac{ik(1 - k/q)}{2\rho\omega}, \qquad (17.47)$$

where

$$q = \sqrt{k^2 - i\omega/\nu}, \tag{17.48}$$

ν is the kinematic viscosity, and the square root is defined such that its real part is positive.

Using the relation $\hat{\dot{H}}_k(\omega) = -i\omega \hat{H}_k(\omega)$, equation (17.46) may be written as

$$\hat{\dot{H}}_k(w) = \Gamma_k(\omega)\hat{F}'_k(\omega) \tag{17.49}$$

where

$$\Gamma_k(\omega) = \frac{\hat{R}_k(\omega)}{1 - \frac{\sigma k^2}{i\omega}\hat{R}_k(\omega)}. \tag{17.50}$$

We have thus obtained an expression for the macroscopic motion of the interface (in frequency-wavenumber space) in terms of the product of the Fourier transforms of a hydrodynamic response function and an arbitrary force.

Now return to the fluctuation-dissipation theorem (17.41). The denominators serve only to establish prefactors. Neglecting them for the moment, we write the proportionality

$$\langle \dot{h}_k(t)\dot{h}^*_k(0)\rangle \propto \dot{H}_k(|t|). \tag{17.51}$$

Fourier transforming both sides, we obtain

$$\langle |\hat{\dot{h}}_k(\omega)|^2\rangle \propto \omega^{-2}[\hat{\dot{H}}_k(\omega) + \hat{\dot{H}}_k(-\omega)]. \tag{17.52}$$

On the left-hand side of the fluctuation-dissipation theorem, the denominator depends on the amplitude of the mean squared fluctuations, whereas on the right-hand side the denominator depends on the amplitude of the macroscopic forcing. If $F'_k(t) = F_0\,\delta(t)$, i.e., a pulse, then the linearity of the problem gives $\dot{H}_k(0) \propto F_0$. Thus the amplitude of the macroscopic force cancels on the right-hand side of equation (17.41). The amplitude of the mean squared fluctuations, on the other hand, may be derived from the equipartition of fluid kinetic energy associated with the interface motion. This can be accomplished by constructing an argument similar to that of Section 17.3, but for $\dot{h}_k$ instead of h_k. The precise expression for the frequency power spectrum is then found to be [17.5,17.7]

$$|\hat{h}_k(\omega)|^2 = \frac{\Theta}{2\pi^2} \frac{k_B T}{\omega^2 L} \,\mathrm{Re}\,\{\Gamma_k(\omega)\}\,, \tag{17.53}$$

where L is the length of the interface.

Figure 17.6 compares the frequency power spectrum $|\hat{h}_k(\omega)|^2$ computed from a simulated interface with the prediction (17.53) for different wavenumbers k. This frequency spectrum was computed from an interface

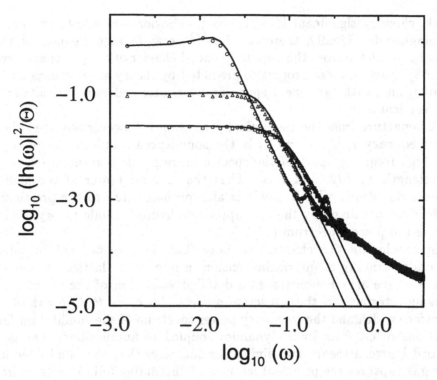

Fig. 17.6.   The frequency power spectrum $|\hat{h}_k(\omega)|^2/\Theta$ for the wavenumbers $k = 2\pi/L$ (circles), $k = 4\pi/L$ (triangles), and $k = 6\pi/L$ (squares), computed from the evolution of an ILG interface of size $L = 32$ over 2^{20} time steps [17.7]. The solid lines show the theoretical prediction (17.53), while the symbols show the corresponding results from simulations. The same effective temperature $k_B T_{\text{eff}}$ was used for all three theoretical curves.

of size $L = 32$ that evolved over $2^{20} \approx 10^6$ time steps. As in Figure 17.5, the surface tension σ required for the theoretical curve was obtained from a Boltzmann approximation; additionally, the viscosity ν was obtained from a Boltzmann estimate of the type described in Chapter 4. Also as in Figure 17.5, an effective temperature $k_B T_{\text{eff}}$ is needed to complete the specification of the theoretical curve. Here $k_B T_{\text{eff}}$ was chosen to minimize the difference between theory and simulation; the particular value that was found, $k_B T_{\text{eff}} = 0.126$, was also used to plot the theoretical curves in Figure 17.5. As in that case, the fit with theory contains no other free parameters.

Theoretically, there are three significant features in the power spectra. First, there is a peak at frequency $\omega = \omega_0(k)$, where $\omega_0(k)$ is given by the capillary-wave dispersion relation (17.30). The peak becomes sharper as $k^2\nu$ becomes smaller. Second, there is a plateau to the left of this peak, showing significant low-frequency behavior over one to two orders of magnitude. Finally, there is a $1/\omega^{7/2}$ decay at high frequencies that may be predicted from the high-frequency behavior of the power spectrum (17.53). Each of these properties predicted by theory are captured by the simulations, with agreement spanning over three orders of magnitude on the vertical axis.

A departure from the theoretical prediction is however observed above the frequency c_s/L, where c_s is the soundspeed. Indeed the peaks in the high-frequency part of the spectra correspond to standing waves of wavelength L, $L/2$, and $L/3$. That the spectral power of the highest frequencies drops off too slowly is also probably due to compressibility effects not predicted by the incompressible hydrodynamic theory used to derive the power spectrum (17.53).

In concluding this chapter, we note that we have not only obtained an understanding of interfacial tension in immiscible lattice-gas models, but we have also demonstrated a detailed validation of the dynamics of moving interfaces in these discrete models. Both the $t^{1/3}$ growth of the interface width and the frequency power spectrum of the equilibrium fluctuations result from hydrodynamics coupled to fluctuations. The good accord between theory and simulation indicates that the immiscible lattice gas captures the principal features of fluctuating fluid interfaces from mesoscopic to macroscopic scales. In particular, the fluctuation dynamics have been probed and verified over a frequency spectrum ranging from the slow behavior of viscous diffusion to the relatively rapid dynamics of capillary wave oscillations and finally to the high-frequency cutoff set by compressibility effects (sound waves). These results show a surprisingly good accord between the predictions of classical fluctuating hydrodynamics and the behavior of a lattice-gas model with time-irreversible microdynamics.

As a consequence of the irreversibility, however, we have had to define an effective temperature $k_B T_{\text{eff}}$ to characterize the fluctuations. The concept of an effective temperature appears to work both in and out of equilibrium. A fundamental understanding of it, however, is unfortunately still beyond our reach.

17.6   Notes

The Boltzmann approximation for ILG surface tension and the associated empirical calculations are from:

17.1 Adler, C., d'Humières, D., and Rothman, D. H. (1994). Surface tension and interface fluctuations in immiscible lattice gases. *J. Physique I France* **4**, 29–46.

Related calculations for interfaces in the liquid-gas model of Chapter 12 are given in Ref. [12.8].

Interface fluctuations and equipartition in classical Ising systems are discussed, for example, in:

17.2 Ma, S.-K. (1985). *Statistical Mechanics* (World Scientific, Singapore).

Experimental studies of fluctuating hydrodynamic interfaces are reported in:

17.3 Bouchiat, M. A. and Meunier, J. (1969). Light scattering from surface waves on carbon dioxide near the critical point. *Phys. Rev. Lett.* **23**, 752–755.

17.4 Bouchiat, M. A. and Meunier, J. (1971). Spectre des fluctuation thermiques de la surface libre d'un liquide simple. *J. Physique France* **32**, 561–571.

The frequency power spectrum (17.53) was apparently first derived in:

17.5 Herpin, J. C. and Meunier, J. (1974). Spectral analysis of light scattered by thermal fluctuations at liquid-vapor interface of CO_2 near its critical point: Measurement of surface tension and viscosity. *J. Physique Colloq.* **35**, 847–859.

Its derivation from fluctuating hydrodynamics, in addition to a discussion of non-equilibrium roughening, dynamical scaling, the hydrodynamic response function (17.47), and comparisons with ILG simulations, are detailed in:

17.6 Flekkøy, E. G. and Rothman, D. H. (1995). Fluctuating fluid interfaces. *Phys. Rev. Lett.* **75**, 260–263.

17.7 Flekkøy, E. G. and Rothman, D. H. (1996). Fluctuating hydrodynamic interfaces: Theory and simulation. *Phys. Rev. E* **53**, 1622–1643.

A derivation of the inviscid capillary wave dispersion relation may be found in:

17.8 Landau, L. D. and Lifshitz, E. M. (1959). *Fluid Mechanics* (Pergamon Press, New York).

Corrections due to viscosity may be found in Ref. [17.7].

As we have emphasized in the text, we have no good understanding of the effective temperature $k_B T_{\text{eff}}$. It is perhaps worth pointing out, however, that semi-detailed balance in the ILG is lacking only at interfacial sites but not in the bulk. We would thus expect that statistical fluctuations in the bulk would correspond to Gibbs states. Since most of the interesting physics occurs at the interface, however, we find this argument incomplete. Moreover, the logarithmic correction to scaling for large interfaces may itself be a correction to $k_B T_{\text{eff}}$. These issues could perhaps be resolved by performing a Green-Kubo analysis of fluctuating hydrodynamics in the ILG, in a spirit similar to:

17.9 Ernst, M. H. and Dufty, J. W. (1990). Hydrodynamics and time correlation functions for cellular automata. *J. Stat. Phys.* **58**, 57–86.

It is interesting to note that the autocorrelation function of the height velocity fluctuations takes the form [17.7]

$$\langle \dot{h}_k(t) \dot{h}_k^*(0) \rangle \sim e^{-k^2 \nu t} (k^2 \nu t)^{-3/2} \qquad (17.54)$$

when $k^2 \nu \ll \omega_0(k)$, where w_0 is given by the capillary wave dispersion relation (17.30). For sufficiently large wavelengths, this correlation function decays as a power law. The velocity-velocity correlation function of a suspended particle in Brownian motion also decays as a power law, as discussed in:

17.10 Alder, B. and Wainwright, T. (1970). Decay of the velocity autocorrelation function. *Phys. Rev. A* **1**, 18–21.

Suspended Brownian particles have been extensively studied by lattice-gas simulations. Two examples of such work are:

17.11 Ladd, A. J. C. and Colvin, M. E. (1988). Application of lattice-gas cellular automata to Brownian motion of solids in suspension. *Phys. Rev. Lett.* **60**, 975–978.

17.12 van der Hoef, M. A., Frenkel, D., and Ladd, A. J. C. (1991). Self-diffusion in colloidal particles in a two-dimensional suspension: Are deviations from Fick's law experimentally observable? *Phys. Rev. Lett.* **67**, 3459–3462.

The precise relation, however, between the correlations found in interface fluctuations and those of Brownian particles remains unexplored.

18

Complex fluids and patterns

One of the greatest virtues of lattice-gas models and lattice-Boltzmann methods is the ease with which they allow one to include microscopic complexity in a model of a fluid. Throughout much of this book we have already considered models of immiscible two-fluid mixtures, the collision rules of which include interactions between particles at neighboring sites. In this chapter, we consider some of the ways in which these ideas may be generalized to create fluids of even greater complexity.

We will cover a fair bit of ground. The models range from toys for the study of pattern formation to methods for the simulation of multiphase flow. Aside from their intrinsic interest as models of complex fluids, these models also serve to indicate how collision rules may be designed to introduce other kinds of microscopic physics into momentum-conserving hydrodynamic models.

We proceed roughly in the order of increasing complexity.

18.1   Stripes and bubbles

We first describe a model that produces some fascinating patterns. Although it conserves momentum, our discussion here is motivated less by fluid mechanics than by pattern formation itself.

The model is a nearly trivial extension of the 2D immiscible lattice gas (ILG) of Chapter 9. In the ILG, red particles and blue particles interact in a way that results in a kind of short-range attraction between particles of the same color. We now introduce a competing long-range repulsion.

We employ the same notation as we used in Section 9.1. Rather than having collisions simply maximize the color flux $\mathbf{q}$ in the direction of the color gradient $\mathbf{f}$ determined from the particles at neighboring sites, we introduce a competing interaction with particles at a distance a. Specifically, we compute the *coarse-grained* gradient

$$\mathbf{g}_a(\mathbf{x}) = \sum_i \mathbf{c}_i \sum_j [r_j(\mathbf{x} + a\mathbf{c}_i) - b_j(\mathbf{x} + a\mathbf{c}_i)] \qquad (18.1)$$

where the parameter a is an integer that defines the scale of coarse-graining. The result of a collision, $r \to r'$, $b \to b'$, is then determined by the r' and b' that maximize

$$\mathbf{q}(r', b') \cdot \left(\frac{\mathbf{f}}{|\mathbf{f}|} - \alpha \frac{\mathbf{g}_a}{|\mathbf{g}_a|} \right), \qquad (18.2)$$

where $\alpha \geq 0$ is a parameter that controls the relative strength of the long-range interaction, and mass, momentum, and color are conserved locally as defined by equations (9.7) and (9.8).

Figure 18.1a shows the result of a simulation in which the the interaction parameter $\alpha = 1.0$, so that the short-range attraction and long-range repulsion are given equal weight. This pattern should be compared to Figure 9.2, which shows phase separation in the ILG. The two simulations are the same except for the addition of the long-range interaction. We see that the basic effect of the long-range interaction is to impede phase separation so that it produces stripes, the width of which is approximately the interaction length a. The stripes have some short-range order but are otherwise of random orientation. Increasing the parameter α increases the correlation length of the stripe orientation.

Figure 18.1b illustrates how the model's behavior depends on the concentration θ of the minority phase. Here α is again set to unity but now $\theta = 1/3$. A pattern of hexagonally ordered bubbles results. Initially the ordering is short range but over time the correlation length becomes much longer. The orientation of the hexagonal order is not related to the underlying lattice but is instead a consequence of the competing interaction.

The results we have just described are for $\alpha = 1.0$. When $\alpha = 0$ the model no longer has a long-range interaction and therefore reduces to the ILG model of Chapter 9. Thus we expect—and indeed observe—a critical value α_c above which the long-range interactions produce stripes or bubbles as in Figures 18.1a and 18.1b, and below which the long-range interaction is insignificant and the usual phase separation ensues. However, when $\alpha = \alpha_c$, that is, neither above nor below the critical point, we find a surprising result, as shown in Figure 18.1c. The pattern is the result of a coexistence of both the bubble phase of Figure 18.1b and the phase-separated state that one expects for $\alpha < \alpha_c$. It probably occurs because α is small enough to prevent the formation of stripes but still large enough to create bubbles from red particles that diffuse into the blue phase and from blue particles dissolved in the red phase.

Aspects of the patterns in Figure 18.1 are present in many physical

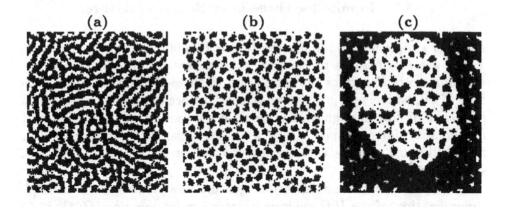

Fig. 18.1. Three simulations of an immiscible lattice-gas model using the competing long-range interaction defined by the maximization of the expression (18.2) [18.1]. In each case the interaction length $a = 5$ and the initial condition was a random mixture. The patterns fluctuate with time, but all three cases are either in equilibrium or close to it. In (a), the interaction strength $\alpha = 1.0$ and the concentration of red (black) is $\theta = 1/2$. The pattern in (b) results from decreasing the concentration to $\theta = 1/3$; all other parameters are the same as in (a). The pattern in (c) results from setting $\alpha = \alpha_c = 0.26$; all other parameters are again the same as in (a). Choosing $\alpha < \alpha_c$ results in phase separation as in Figure 9.2.

systems, such as Langmuir monolayers and diblock copolymers, to name just two. Though these systems themselves result from competing interactions, the interaction rule described here is motivated more by simplicity than by a particular physical prescription. It is nevertheless interesting to ask what kind of physical system would behave like the model we just described. For example, in our discussion of phase separation in Section 16.1, we considered the Ginzburg-Landau free-energy functional (16.2) to be a mesoscopic thermal analog of the ILG. Proceeding along similar lines, the competing-interaction model could be described by the modified free-energy functional

$$F[\phi] = \int \mathrm{d}\mathbf{x} \left[\frac{\xi^2}{2} \left(|\nabla\phi|^2 - \alpha |\langle\nabla\phi\rangle_a|^2 \right) + f_e(\phi) \right], \qquad (18.3)$$

where $f_e(\phi)$ is the same expansion of the order parameter ϕ given by equation (16.1) and $\langle\cdot\rangle_a$ implies a coarse-grained average over an area of linear dimension a. A study of the equilibrium properties of the thermodynamic system defined by the free energy (18.3) indeed predicts aspects of the striped and bubble patterns of Figure 18.1 [18.1].

18.2   Immiscible three-fluid Boolean mixtures

Another simple way to modify the ILG is to add a third species of fluid. There are two reasons to consider such an extension. First, it makes available to lattice gases the simulation of three-phase flow, such as the flow of oil, water, and gas. Three-phase flow occurs, for example, in the problems of flow through porous media discussed in Chapter 13. Second, non-equilibrium phenomena such as phase separation produce qualitatively different patterns with three fluids than with two, as we shall soon see.

An immiscible three-fluid lattice gas may be created from a simple generalization of the ILG collision rules given by equations (9.4), (9.5), and (9.6). We now represent three species of fluids by the Boolean variables $(n_i^j)_{0 \le i \le 6}$, where the superscript $j = 1, 2, 3$ is used to index the fluid species, say, red, green, and blue, and an exclusion rule holds such that for any velocity i, at most one n_i^j may equal unity. On the two-dimensional triangular lattice, the flux of species j is then

$$\mathbf{q}_j[n_0^j(\mathbf{x}), \dots, n_6^j(\mathbf{x})] = \sum_{i=1}^{6} \mathbf{c}_i n_i^j(\mathbf{x}), \qquad j = 1, 2, 3, \qquad (18.4)$$

while the local gradient of the jth species is proportional to

$$\mathbf{f}_j = \sum_k \mathbf{c}_k \sum_i n_i^j(\mathbf{x} + \mathbf{c}_k), \qquad j = 1, 2, 3. \qquad (18.5)$$

The result of a collision, $(n_i^j) \to (n_i'^j)$, is then the choice of configuration that maximizes the weighted sum

$$\sum_j \alpha_j \mathbf{f}_j \cdot \mathbf{q}_j(n_0'^j, \dots, n_6'^j), \qquad (18.6)$$

subject to conservation of each species,

$$\sum_i n_i'^j = \sum_i n_i^j, \qquad j = 1, 2, 3, \qquad (18.7)$$

and conservation of total momentum,

$$\sum_i \sum_j \mathbf{c}_i n_i'^j = \sum_i \sum_j \mathbf{c}_i n_i^j. \qquad (18.8)$$

The coefficients α_j are chosen to set the three surface tensions, σ_{12}, σ_{13}, and σ_{23}. If the α_j's are all equal, then so are the surface tensions, and three-phase contact lines (or points in two dimensions) are not only stable, but act as the point of contact for three interfaces, each making an angle of $2\pi/3$ with respect to the others. If, on the other hand, the α_j are

(a)   (b)

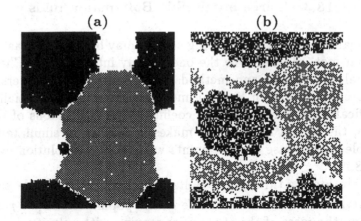

Fig. 18.2.   Phase separation in two different simulations of a three-fluid immiscible lattice gas model [18.2]. In both cases the initial condition is a random mixture of three fluids, here indicated by the colors black, grey, and white, each with equal volume fraction. Boundary conditions are periodic in both directions. The left-hand side (a) is the result of nearly complete phase separation for a simulation in which all three surface tension coefficients are equal. The resulting three-phase contact points are therefore stable. On the right-hand side (b) is the result of partial phase separation in which the three surface tensions satisfy the inequality (18.9). Three-phase contact points are thus unstable and the white phase is always between the grey phase and the black phase. The many white dots in the black and grey phases are due to dissolution of the white phase.

chosen so that the surface tensions are such that, say,

$$\sigma_{13} + \sigma_{12} < \sigma_{23}, \tag{18.9}$$

i.e., the sum of two surface tensions is less than the third, then three-phase contact points are not stable, and the mixture is in equilibrium only when species 2 and 3 are everywhere separated by species 1.

Figure 18.2 shows two examples of phase separation with the three-phase Boolean ILG. In one case, each α_j and thus all three surface tensions are equal. As expected, all three possible pairs of fluids create interfaces that form angles of approximately 120 degrees with respect to each other. The second case is the result of a simulation with the parameter choice $\alpha_1 = -0.5$ and $\alpha_2 = \alpha_3 = 1$, chosen to satisfy the inequality (18.9) and thus make three-phase contact points unstable. As seen in the figure, the white phase has inserted itself everywhere between the black and grey phases. As a consequence of the small surface tension at interfaces formed with the white fluid, there is also considerable dissolution of the white particles into the black and grey fluids.

18.3 Three immiscible Boltzmann fluids

The Boolean model of the preceding section may be expressed as a lattice-Boltzmann method similar to the one already introduced in Chapter 10. As is usual with Boltzmann methods, the choices of parameters may be made more freely than in Boolean models. This can be especially useful for specification of surface tension coefficients in simulations of three fluids. Also, the lack of fluctuations makes it possible to simulate the case of unstable three-phase contact points without the dissolution evident in Figure 18.2.

The evolution of a three-phase Boltzmann method follows the same four steps described in Section 10.1. We define populations $\Phi_i^{(j)}$, $j = 1, 2, 3$, designating the mass of the jth species moving with velocity c_i. We take N_i to be the usual "colorblind" population variable such that

$$N_i = \sum_j \Phi_i^{(j)}. \tag{18.10}$$

The first step is to apply the collision equation (10.1) such that $N_i \to N_i'$.

The second step is, as before, responsible for the surface tension. We calculate the three gradient vectors

$$\mathbf{f}_j = \sum_k \mathbf{c}_k \sum_i \Phi_i^{(j)}(\mathbf{x} + \mathbf{c}_k), \qquad j = 1, 2, 3 \tag{18.11}$$

and perturb the populations N_i' such that the flux of the jth species is parallel to the gradient of the jth species. Thus, analogous to equation (10.3), we have

$$N_i''(\mathbf{x}, t) = N_i'(\mathbf{x}, t) + \sum_j A_j |\mathbf{f}(\mathbf{x}, t)| \left[\frac{(\mathbf{c}_i \cdot \mathbf{f}_j)^2}{\mathbf{f}_j \cdot \mathbf{f}_j} - \frac{1}{2} \right], \tag{18.12}$$

where the A_j are parameters used to set the three surface tensions.

The third step is to recolor the mass N_i'' to minimize the diffusion of each species. We choose $\Phi_i''^{(j)}$ such that

$$\sum_i \Phi_i''^{(j)} \mathbf{c}_i \cdot \mathbf{f} \qquad j = 1, 2, 3 \tag{18.13}$$

is maximized for each species j, subject to the conservation of the mass moving in each direction $\mathbf{c}_i$,

$$\sum_j \Phi_i''^{(j)}(\mathbf{x}, t) = N_i''(\mathbf{x}, t). \tag{18.14}$$

The last step is then free propagation:

$$\Phi_i^{(j)}(\mathbf{x} + \mathbf{c}_i, t + 1) = \Phi_i''^{(j)}(\mathbf{x}, t). \tag{18.15}$$

(a)   (b)

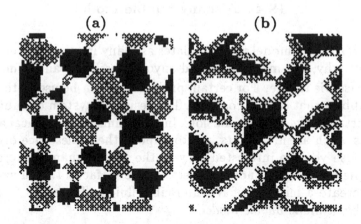

Fig. 18.3.   Three-fluid phase separation with a lattice-Boltzmann method [18.3].
Both simulations should be compared with their respective Boolean equivalents
in Figure 18.2. As in that figure, the result on the left (a) was computed with
equal surface tensions, whereas the simulation on the right (b) was performed
with unequal surface tensions such that triple points are unstable and the inequality (18.9) is satisfied. Here it is the grey fluid that inserts itself everywhere
between the other two fluids in the unstable case As a consequence of the lack
of fluctuations, there is no dissolution of the grey phase into the others.

The surface tensions σ_{ij} of the interfaces separating the ith and jth
species may be calculated as in Chapter 10. Expressing the result in
terms of the two-fluid surface tension σ derived in Section 10.2, one finds

$$\sigma_{ij} = \left(\frac{A_i + A_j}{2} \right) \frac{\sigma}{A}, \tag{18.16}$$

where σ/A is to be read from equations (10.26) and (10.28) for the cases
of the hexagonal and FCHC lattices, respectively.

Examples of phase separation in the three-fluid Boltzmann model are
shown in Figure 18.3. As in Figure 18.2, cases with both stable and unstable triple points are shown. Though the qualitative behavior of the
Boltzmann method is the same as in the Boolean model, it is interesting
to note two differences due to the lack of fluctuations in the Boltzmann
method. First, we see that in the case of stable triple points, phase separation with the Boolean model has resulted in fewer and bigger bubbles
than with the Boltzmann method, even though both systems have reached
their equilibrium configuration. Secondly, we see in the case of unstable
triple points that the Boltzmann method does not result in any diffusion
of the interstitial species into the others, as is strongly evident with the
Boolean model.

18.4   A many-bubble model

We now describe a model that simulates as many immiscible fluids as you want. But why, you may ask, would anyone possibly want to do that?

The reason is simple. For certain problems, it can be useful to simulate many bubbles that *cannot* coalesce. The resulting system of bubbles then acts in certain ways like an emulsion. In real emulsions, chemical agents on interfaces, such as surfactants, act to impede the coalescence of bubbles. But suppose you are interested only in the hydrodynamic interaction of deformable bubbles, but not the effects of the surfactant. You would then want to create a mixture of N immiscible fluids! Taking N to be large, we call this a *many-bubble model*.

A Boolean version of such a model may be constructed without much difficulty in 2D because the number of different particle species that may collide at a site is small. Indeed, for the case of one rest particle on the hexagonal lattice, no more than seven different fluid species may enter a collision.

The idea is to index each particle on the lattice by an integer species code between 1 and $N + 1$. We choose species 1 to be the interstitial or suspending fluid. Then, at each site and before each collision, the two locally most numerous species other than the interstitial fluid are temporarily labeled species 2 and 3. Now assume that there are no other particle types in the collision. The collision may then be performed using the immiscible three-phase collision rules described in Section 18.2, choosing the weights α_j so that three-phase contact points are unstable. After the collision, the species that were temporarily labeled "2" and "3" are then remapped back to their original integer codes. Using such a rule, species 2 through $N + 1$ form N bubbles in a sea of species 1.

What happens if there are more than two non-interstitial species entering a collision? If the model is initialized as N bubbles, such collisions will happen only rarely, but they will occur. One solution is to leave these extra particles in place and let them be "spectators" to the collision. Another is to group them all (there can be no more than five) into a species temporarily labeled "4" and to then generalize the three-phase rule of Section 18.2 to four fluids. Other solutions may also be imagined—one need only insure that mass, momentum, and particle type are conserved.

Figure 18.4 shows the result of simulating the many-bubble model without external forcing. Each bubble, shown in black, is a distinct fluid species; thus the bubbles may never coalesce. Since there is no forcing, each bubble undergoes a random walk due to the statistical noise of the lattice gas. The model thus simulates the collective hydrodynamic interactions of N deformable bodies in Brownian motion.

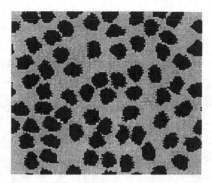

Fig. 18.4. Equilibrium configuration in a simulation of the many-bubble model [18.5]. The lattice is 128 × 128 and each bubble has a radius of about five lattice units. The random placement of each bubble results from Brownian motion due to the statistical fluctuations of the lattice gas.

Figure 18.5 shows a simulation of *two-component sedimentation* performed with the the many-bubble model. Here there are more than one thousand bubbles. Half of them are considered "hot" and are colored white, while the other half are considered "cold" and are colored grey. (Heat does not diffuse.) The white bubbles are forced upward, the grey bubbles are forced downward, and the interstitial black fluid is unforced. In other words, hot bubbles rise and cold bubbles fall. The simulation was initialized with a random mixture of hot and cold bubbles. For a sufficiently high volume fraction of bubbles, however, the mixture is unstable and the bubbles segregate into regions composed primarily of white bubbles or grey bubbles, as we see here. Such instabilities in two component sedimentation are known from experiments and, to a lesser extent, from theory [18.6].

18.5 Microemulsions

Our final foray into complex fluids is an attempt to simulate the hydrodynamics and statistical mechanics of an immiscible mixture to which an *amphiphile* species has been added. The corresponding real-world problem is an oil-water mixture in the presence of a surfactant. The amphiphile or surfactant molecules are *polar*, that is, they have a hydrophilic head and a hydrophobic tail. They thus prefer to be situated at oil-water interfaces. As surfactant is added to a mixture, the equilibrium surface area of the mixture tends to grow. Stable structures such as a mixture of oil droplets in water or a sponge-like bicontinuous phase can then exist in equilibrium. These complex fluid structures are known as *microemulsions*.

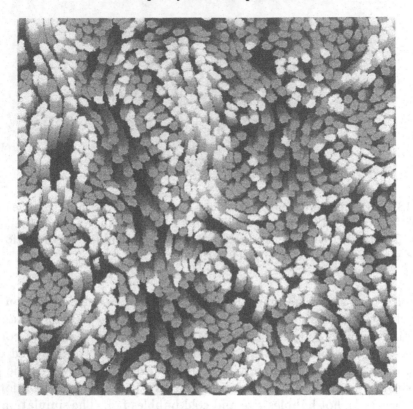

Fig. 18.5. Simulation of two-component sedimentation using the many-bubble model [18.5]. The lattice size is 512×512. Positively buoyant bubbles are white and negatively buoyant bubbles are grey; the interstitial fluid is black. There are 1024 bubbles, each with a radius of about five lattice units, encompassing a total volume fraction of 0.4. The initial condition was a random mixture of white and grey bubbles. Here, 8500 time steps later, structures similar to fingers, plumes, and columns have organized as a consequence of the bubble motions. The recent movement of the individual bubbles prior to this snapshot is indicated by a reverse fade-out: the more distant in time, the more pale is the trail left by the bubble.

The idea, due to Boghosian, Coveney, and Emerton [18.7], is to add to the ILG a third species with a vector degree of freedom, and have it collide with the two other species in a way that encourages its placement at interfaces. Thus in addition to red particles r_i and blue particles b_i, we have also an amphiphile species denoted by the unit vector $\mathbf{a}_i$. We may think of $\mathbf{a}_i$ as a Boolean particle a_i of unit mass that carries an angular degree of freedom in the direction $\mathbf{c}_i$. The angular degree of freedom may be either real-valued or limited to a discrete set of directions (for example, those of the lattice). Once again we have an exclusion rule, so that at most one element of the set (r_i, b_i, a_i) may be non-zero at a particular site and time step. Also, the microdynamical evolution of the model takes a form similar to that of the ILG. It is given by the equation (9.3) for the red and blue species and, analogously,

$$\mathbf{a}_i(\mathbf{x} + \mathbf{c}_i, t + 1) = \mathbf{a}_i'(\mathbf{x}, t), \tag{18.17}$$

for the amphiphiles.

In addition to the ILG rule that tends to align the color flux $\mathbf{q}$ with the color gradient $\mathbf{f}$ as detailed in Section 9.1, we also wish to specify collision rules that govern the coupling of the amphiphile vectors $\mathbf{a}_i$ to the red and blue fluids. We first describe what appears to be an essential aspect of modeling a microemulsion—a rule that chooses $\mathbf{a}_i \rightarrow \mathbf{a}_i'$ such that the post-collision amphiphile vectors $\mathbf{a}_i'$ tend to point in the direction of the color gradient $\mathbf{f}$ in addition to moving in a direction that encourages their placement at interfaces.

We shall need the following notation, written for the case of the hexagonal lattice with one rest particle. The total amphiphile population at a site is given by

$$\overline{\mathbf{a}}(\mathbf{x}) = \sum_i \mathbf{a}_i. \tag{18.18}$$

In analogy with the color flux $\mathbf{q}$ given by equation (9.5), we define the *amphiphile flux tensor*

$$Q_{\alpha\beta}[\mathbf{a}_0(\mathbf{x}), \ldots, \mathbf{a}_6(\mathbf{x})] = \sum_i c_{i\alpha} a_{i\beta}(\mathbf{x}), \tag{18.19}$$

which gives the β-component of the α-flux of amphiphile vectors at the site located at $\mathbf{x}$. We also define the *second-order color gradient*

$$F_{\alpha\beta}(\mathbf{x}) = \sum_i c_{i\alpha} f_{\beta}(\mathbf{x} + \mathbf{c}_i), \tag{18.20}$$

where f_{β} is the β-component of the color gradient $\mathbf{f}$ given by equation (9.6). Collisions $r_i \rightarrow r_i'$, $b_i \rightarrow b_i'$, and $\mathbf{a}_i \rightarrow \mathbf{a}_i'$ may then be specified such that

$$(\mathbf{q} + A\overline{\mathbf{a}}') \cdot \mathbf{f} + A Q_{\alpha\beta} F_{\alpha\beta} \tag{18.21}$$

is maximized, subject to the conservation of species,

$$\sum_i r_i' = \sum_i r_i, \qquad \sum_i b_i' = \sum_i b_i, \qquad \sum_i a_i' = \sum_i a_i, \tag{18.22}$$

and the conservation of momentum,

$$\sum_i \mathbf{c}_i(r_i' + b_i' + a_i') = \sum_i \mathbf{c}_i(r_i + b_i + a_i). \tag{18.23}$$

In expression (18.21), $\mathbf{q}$ depends on the primed populations r_i' and b_i' while $\bar{\mathbf{a}}'$ and $Q_{\alpha\beta}$ depend on a_i'. The parameter A is chosen to determine the relative importance of the amphiphile species in the collisions. Note that when $A = 0$ the collision rules reduce to those of the ILG.

The two terms weighted by A have the following interpretation. The first simply encourages the amphiphile vectors to align themselves with $\mathbf{f}$. In other words, the hydrophilic head tends to point towards the water and the hydrophobic tail towards the oil. The second term is more subtle. It tends to send the amphiphile flux $Q_{\alpha\beta}$ towards locations where the gradient of $\mathbf{f}$ is greatest and away from locations where it is smallest. Hence this term causes the amphiphiles to situate themselves at red-blue interfaces.

We are not necessarily done with our specification of a microemulsion, however. In principle, we should also account for interactions among the amphiphiles themselves, as well as the amphiphilic contribution to the color field $\mathbf{f}$. To include these interactions, we need terms analogous to $\mathbf{f}$ and $F_{\alpha\beta}$. We first define the scalar field

$$S(\mathbf{x}) = \sum_i \mathbf{c}_i \cdot \left(\sum_j \mathbf{a}_j(\mathbf{x} + \mathbf{c}_i) \right), \qquad (18.24)$$

which may be thought of as the divergence of the local amphiphile field. We then define the *amphiphile gradient*

$$\mathbf{g}(\mathbf{x}) = \sum_i \mathbf{c}_i S(\mathbf{x} + \mathbf{c}_i) \qquad (18.25)$$

and the *second-order amphiphile gradient*

$$G_{\alpha\beta}(\mathbf{x}) = \sum_i c_{i\alpha} g_\beta(\mathbf{x} + \mathbf{c}_i). \qquad (18.26)$$

The modified collision rule would then choose r_i', b_i', and a_i' to maximize

$$(\mathbf{q} + A\bar{\mathbf{a}}') \cdot (\mathbf{f} + \mathbf{g}) + A Q_{\alpha\beta}(F_{\alpha\beta} + G_{\alpha\beta}), \qquad (18.27)$$

rather than the simpler form (18.21).

Instead of the one-parameter form (18.27), Boghosian et al. [18.7] suggest that it be written as

$$-\Delta E \equiv A_1 \mathbf{q} \cdot \mathbf{f} + A_2 \mathbf{q} \cdot \mathbf{g} + A_3(\bar{\mathbf{a}}' \cdot \mathbf{f} + Q_{\alpha\beta} F_{\alpha\beta}) + A_4(\bar{\mathbf{a}}' \cdot \mathbf{g} + Q_{\alpha\beta} G_{\alpha\beta}). \qquad (18.28)$$

Here A_1 is a parameter that controls the surface tension, A_2 controls the tendency of amphiphile vectors to bend around bubbles, A_3 has the same meaning as the unsubscripted A of equation (18.21), and A_4 controls the propensity of amphiphile particles to align and move parallel to neighboring amphiphiles. Moreover, Boghosian et al. argue that (18.28) is

(a)   (b)

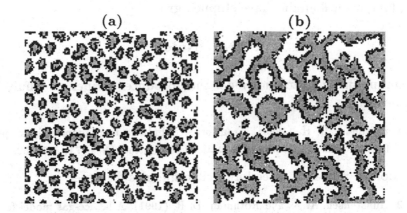

Fig. 18.6. Snapshots from the equilibrium state of two simulations of a microemulsion [18.7]. The initial condition in both cases was a random mixture of red (white), blue (grey), and amphiphile (black) particles such that the total particle density was approximately the same. In (a), the red-to-blue-to-amphiphile ratio of densities is 3 : 0.7 : 1.9, while in (b) it is 3 : 2.25 : 3. Note that the amphiphile species is located almost always at the interfaces. Since it acts to impede coalescence, the domains in both cases can no longer grow in size. The pattern in (a) is analogous to an oil-in-water microemulsion, while the pattern in (b) is a 2D version of a sponge-like bicontinuous microemulsion.

analogous to a change in interaction energy, thus motivating its definition as $-\Delta E$. Then, rather than minimizing ΔE, Boghosian et al. choose r_i', b_i', and $\mathbf{a}_i'$ probabilistically according to Boltzmann weights $\exp\{-\Delta E\}$.

Figure 18.6 shows the result of two simulations of the microemulsion model performed by Boghosian et al. Both simulations used the parameters

$$A_1 = 1.0, \qquad A_2 = 0.05, \qquad A_3 = 8.0, \qquad A_4 = 0.5, \qquad (18.29)$$

which implies that the dynamics were not too different from those that would have resulted from a rule which maximized expression (18.21) using $A = 8$. The two simulations were both started with a random mixture of red, blue, and amphiphile particles such that the total particle density was approximately the same. The relative concentrations of the three species, however, differ. In the simulations, the mixture first starts to phase separate. However, the amphiphile particles cluster at the red-blue interfaces and make it difficult for bubbles to coalesce. Thus, rather than achieving a full phase separation as in Figure 9.2, the structures in the microemulsion eventually reach a characteristic size and grow no longer. The two snapshots in Figure 18.6 are indeed taken in the equilibrium state, long after domains have stopped growing.

18.6   Notes

The competing interaction model that produces stripes and bubbles is
described in:

18.1 Rothman, D. H. (1993). From ordered bubbles to random stripes: pattern
formation in a hydrodynamic lattice gas. *J. Stat. Phys.* **71**, 641–652.

The Boolean model of three immiscible fluids is discussed in:

18.2 Gunstensen, A. K. and Rothman, D. H. (1991). A lattice-gas model for three
immiscible fluids. *Physica D* **47**, 47–52.

The lattice-Boltzmann form of this three-fluid model was introduced in
Chapters 4 and 5 of:

18.3 Gunstensen, A. K. (1992). *Lattice-Boltmann Studies of Multiphase Flow
Through Porous Media*. Ph. D. thesis, Massachusetts Institute of Technology.

The many-bubble model is discussed in:

18.4 Rothman, D. H. (1992). Simple models of complex fluids. In Mareschal, M.
and Holian, B., editors, *Microscopic Simulations of Complex Hydrodynamics*,
pages 221-238 (Plenum Press, New York).

18.5 Rothman, D. H. and Kadanoff, L. P. (1994). Bubble, bubble, boil and trouble.
Computers in Physics **8**, 199–204.

The second of these references discusses an application to Rayleigh-
Bénard convection in addition to the problem of two-component sedimentation. Theoretical and experimental aspects of two-component sedimentation are discussed in:

18.6 Batchelor, G. K. and Janse van Rensburg, R. W. (1986). Structure formation
in bidisperse sedimentation. *J. Fluid Mech.* **166**, 379–407.

Lattice-gas models of microemulsions were introduced in:

18.7 Boghosian, B. M., Coveney, P. V., and Emerton, A. N. (1996). A lattice-gas
model of microemulsions. *Proc. Roy. Soc. Lond. A* **452**, 1221–1250.

Appendix A
Tensor symmetry

The purpose of this appendix is to give several related geometrical facts that are useful for the derivation of the symmetry properties of lattice gas hydrodynamics. We start with a review of general facts about tensors and symmetries. Symmetries are of two kinds: space symmetries transform ordinary space, while tensors have additional symmetries under permutation of the indices that must be discussed separately. This discussion is followed with a discussion of specific lattice symmetries. We prove some important symmetry properties for tensors associated with the lattices considered in this book. These symmetry properties are needed for the derivation of both inviscid and viscous hydrodynamics. The computation of specific tensors based on the velocity vectors is also addressed.

A.1 Space symmetry: isometry groups

An isometry (or an orthogonal transformation, or a congruent transformation) is a transformation of space that leaves distances invariant. We will need to consider only transformations leaving the origin invariant. They may be defined as an operator with matrix $\mathbf{R}$ acting on vectors $\mathbf{x}$. The new coordinates after a rotation will be

$$x'_\alpha = R_{\alpha\beta}x_\beta \tag{A.1}$$

where we have used *Einstein's notation* in which a summation is implied on all repeated indices. From the invariance of the Euclidean norm $|\mathbf{x}|$ one obtains the standard linear algebra result $R_{\alpha\beta}R_{\alpha\gamma} = \delta_{\beta\gamma}$. Thus we find the inverse of $\mathbf{R}$ by $(R^{-1})_{\alpha\beta} = R_{\beta\alpha}$ which means that the matrix's inverse is its transpose. In another form

$$\mathbf{R}\mathbf{R}^T = \mathbf{R}^T\mathbf{R} = \mathbf{I} \tag{A.2}$$

where $\mathbf{I}$ is the identity matrix. The characterization of all isometries is a classic piece of geometry [A.2]. One finds *reflections about a plane* and *rotations*. Isometries of D-dimensional space form a group classically called the *orthogonal* group $O(D)$, while the set of rotations only forms the group $SO(D)$, i.e. the *special orthogonal group*.

It is interesting to consider the finite subgroups of $O(D)$, especially since the finite groups of isometries are the symmetry groups of regular polygons, polyhedra etc. In 2D we have the groups that leave polygons invariant. For instance the symmetry group of the triangle contains six isometries: three reflections about the axes joining the center to the mid point of an edge, two rotations of the form

$$\mathbf{R}_{2\pi/3} = \begin{pmatrix} -\frac{1}{2} & -\frac{\sqrt{3}}{2} \\ \frac{\sqrt{3}}{2} & -\frac{1}{2} \end{pmatrix} \tag{A.3}$$

and $\mathbf{R}_{-2\pi/3} = \mathbf{R}_{2\pi/3}^{-1}$, and the identity. The square has eight isometries and the hexagon twelve. Generally a polygon with n edges has $2n$ isometries.

The symmetry groups of the square, the cube or the hypercube have an interesting structure that will be shown in Appendix B to be very useful. Consider a cube with coordinates $\mathbf{x} = (\epsilon_1, \epsilon_2, \epsilon_3)$ where each ϵ_i is either $+1$ or -1. The cube is invariant by the *parity transformations* $\mathbf{P}_i$ which change coordinate x_i into $-x_i$. For instance

$$\mathbf{P}_1 = \begin{pmatrix} -1 & 0 & 0 \\ 0 & 1 & 0 \\ 0 & 0 & 1 \end{pmatrix} \tag{A.4}$$

which generalizes easily to any dimension. The matrices $\mathbf{P}_i$ commute and generate a group $\mathcal{R}$ of parity transformations of order[*] eight in 3D. Any element of $\mathcal{R}$ may be written uniquely in the form

$$\begin{pmatrix} \mathbf{I} \\ \mathbf{P}_1 \end{pmatrix} \begin{pmatrix} \mathbf{I} \\ \mathbf{P}_2 \end{pmatrix} \begin{pmatrix} \mathbf{I} \\ \mathbf{P}_3 \end{pmatrix} \tag{A.5}$$

where the above notation means a product of three matrices, each chosen among two possible matrices. The cube is also invariant by transformations $\mathbf{P}_{ij}$ that exchange two coordinates x_i and x_j. For instance

$$\mathbf{P}_{12} = \begin{pmatrix} 0 & 1 & 0 \\ 1 & 0 & 0 \\ 0 & 0 & 1 \end{pmatrix}. \tag{A.6}$$

---

[*] Recall that the order of a group is the number of its elements.

These transformations generate the group $\mathcal{T}_\sigma$ of *coordinate permutations*. Its structure is well known from elementary group theory. There are six possible permutations of three coordinates, so in 3D $\mathcal{T}_\sigma$ has six elements. Together $\mathcal{T}_\sigma$ and $\mathcal{R}$ generate the *octahedral group* of cube transformations. To see why we have to use the *orbit stabiliser* theorem which is stated in Appendix B and which results in the fact any isometry of the cube may be written uniquely as the product of an isometry in $\mathcal{T}_\sigma$ and an isometry in $\mathcal{R}$. Thus there are $8 \times 6 = 48$ isometries in the octahedral group.

A.2 Tensor symmetries

Tensors are multiply-indexed objects. A *scalar* (e.g. the temperature) has no indices. A vector v_α has one index and is a rank one tensor. A matrix $M_{\alpha\beta}$ is a rank two tensor. Generally an object with k indices is a rank k tensor $T_{\alpha_1\cdots\alpha_k}$. When space is transformed by an isometry $\mathbf{R}$ the coordinates of a tensor change accordingly.[†] The physical meaning of this transformation is the following. Suppose the tensor measures some property of the medium, such as its elasticity or viscosity. If we change the coordinate system, and measure the same property again, the tensor that describes it is changed in the way above. It amounts to the same thing to rotate the system of coordinates or to rotate the medium. If we cannot tell the difference between the original medium and the rotated medium then the measured tensor should be the same: it should be *invariant*. A medium is isotropic if it is invariant by all isometries, i.e. all rotations and reflections.[‡]

We now recall the rule for transformations of tensors under a space symmetry or transformation. Scalars are unchanged by these transformations. A vector is transformed by $v'_\alpha = R_{\alpha\beta}v_\beta$. More generally, a rank k tensor transforms as

$$T'_{\alpha_1\cdots\alpha_k} = T_{\beta_1\cdots\beta_k} R_{\alpha_1\beta_1} \cdots R_{\alpha_k\beta_k}. \tag{A.7}$$

For instance, for a tensor of order two in $D = 2$ dimensions,

$$T'_{11} = R_{11}R_{11}T_{11} + R_{11}R_{12}T_{12} + R_{11}R_{12}T_{21} + R_{12}R_{12}T_{22}. \tag{A.8}$$

The action of isometries in the octahedral group is particularly simple. For instance, the action of $\mathbf{P}_1$ is immediately found from equation (A.8) to be $T'_{12} = -T_{12}$. It will be useful to remember the more general result that $\mathbf{P}_1$ will multiply $T_{\alpha_1\cdots\alpha_k}$ by $(-1)^n$ where n is the number of times the

---

[†] As we consider only orthogonal transformations of space we do not need to distinguish between covariant and contravariant indices, as is sometimes done in tensor theory.

[‡] One may imagine some special media which are invariant by all rotations but not by reflections. Such *chiral* media will not be encountered in this book.

index 1 appears in $\alpha\beta\gamma\delta$. The action of $\mathbf{P}_{12}$ is to transpose coordinates one and two. For instance $T'_{1222} = T_{2111}$, and more generally all 1's are changed into 2's and all 2's are changed into 1's. As further warm-up, it is interesting to consider the effect of a $2\pi/3$ rotation. The corresponding rotation matrix is $R_{2\pi/3}$. Using equation (A.3) and (A.8) we find

$$T'_{11} = \frac{1}{4}T_{11} + \frac{\sqrt{3}}{4}(T_{12} + T_{21}) + \frac{3}{4}T_{22}. \tag{A.9}$$

This will be useful below for hexagonal symmetry.

Invariant or isotropic media are characterized by invariant tensors. The tensor $\mathbf{T}$ is invariant by the transformation (A.7) if $\mathbf{T}' = \mathbf{T}$. For instance, matrices $\mathbf{M}$ are invariant if

$$M_{\alpha\beta} = M_{\alpha'\beta'}R_{\alpha\alpha'}R_{\beta\beta'}. \tag{A.10}$$

Notice that for an isometry, from the property of inverses in equation (A.1) we find that equation (A.10) may be written $\mathbf{M} = \mathbf{RMR}^{-1}$ which is equivalent to $\mathbf{MR} = \mathbf{RM}$. Thus the matrix $\mathbf{M}$ *commutes* with the isometry $\mathbf{R}$.

In addition to space symmetry we have special symmetries for tensors: tensors may be invariant under *permutation of the indices*. The idea of permutation of the indices should be considered carefully, and not confused with invariance under $\mathcal{T}_\sigma$. A tensor is *symmetric under permutations of the indices* if the order in which its indices are listed may be rearranged. This should not be confused with *invariance* under the group $\mathcal{T}_\sigma$ of isometries. For instance if a tensor $\mathbf{T}$ is invariant under permutation of the indices, then $T_{1222} = T_{2122} = T_{2212} = T_{2221}$. On the other hand, if $\mathbf{T}$ is invariant under permutation of the coordinates then $T_{1222} = T_{2111} = T_{3444} = \cdots$ with many other possibilities.

There are two instances where we need to consider permutation of the indices. Consider again equation (2.52),

$$\Pi^{(0)}_{\alpha\beta} = p_0\delta_{\alpha\beta} + T_{\alpha\beta\gamma\delta}u_\gamma u_\delta + \mathcal{O}(u^4), \tag{A.11}$$

where we have renamed the tensor $\mathbf{T}$. The tensor $T_{\alpha\beta\gamma\delta}$ must have the general symmetry properties of the medium (isotropy for an isotropic fluid). In addition, we can clearly assume that indices γ and δ may be exchanged. If that were not the case, it suffices to notice that

$$T_{\alpha\beta\gamma\delta}u_\gamma u_\delta = \frac{1}{2}(T_{\alpha\beta\gamma\delta} + T_{\alpha\beta\delta\gamma})u_\gamma u_\delta. \tag{A.12}$$

Thus upon redefining $\mathbf{T}$ we find the desired symmetry. Moreover, from the definition of $\Pi^{(0)}$ we have $\Pi^{(0)}_{\alpha\beta} = \Pi^{(0)}_{\beta\alpha}$ so the indices α and β may also be exchanged.

The second instance is the constitutive relation for viscosity, equation (2.22), which reads

$$\Pi_{\alpha\beta}^{\text{visc}} = -\tilde{\mu}_{\alpha\beta\gamma\delta}E_{\gamma\delta}. \qquad (A.13)$$

Again the symmetry of the strain rate tensor $E_{\gamma\delta} = \partial_\gamma u_\delta + \partial_\delta u_\gamma$ ensures the symmetry of γ and δ. There is however a question. We may wonder why only the symmetric form $E_{\gamma\delta}$ appears and not the antisymmetric rotation rate $\Omega_{\gamma\delta} = \partial_\gamma u_\delta - \partial_\delta u_\gamma$. In real fluids, the absence of the antisymmetric part derives from the following reasoning. A continuous medium undergoing a uniform translation and a solid body rotation has a velocity field

$$u = -\omega y + u_0 \qquad (A.14)$$
$$v = \omega x + v_0 \qquad (A.15)$$

where u_0, v_0 and ω are constants. This medium undergoes no dissipation. Thus when the strain tensor $\partial_\alpha u_\beta + \partial_\beta u_\alpha = 0$, the viscous stress, and the dissipation, are zero. Thus

$$\mu_{\alpha\beta\gamma\delta} = \mu_{\beta\alpha\delta\gamma}. \qquad (A.16)$$

But in lattice gases, this reasoning does not hold. For the same reason that we do not have Galilean invariance (see Chapter 2), we also do not have invariance under solid body rotations. There is however a saving grace. As shown in Chapter 2 and also below, an isotropic tensor that is symmetric under exchange of the *first* pair of indices is also invariant under exchange of the *second* pair. The viscous tensor $\Pi_{\alpha\beta}^{\text{visc}}$ turns out most of the time to be invariant under permutation of α and β. Thus we have the desired symmetry after all.

Symmetry under permutation of α and β is a specific property of the lattice-gas models introduced in most of this book, but not of all of them. In particular those models in which collisions have a non-zero impact parameter (for instance liquid-gas models in which transverse momentum is exchanged in addition to those exchanges described in Chapter 12) may create torque in the system and thus have a non-symmetric viscous tensor $\Pi_{\alpha\beta}^{\text{visc}}$!

A.3 Isotropic tensors

The *isotropic* tensors (invariant by all rotations and reflections) are especially interesting to characterize. Scalars are always isotropic. Vectors are isotropic only if they vanish. In rank two the isotropic tensors (i.e. matrices) are proportional to $I_{\alpha\beta} = \delta_{\alpha\beta}$. In rank three we find again the null tensor, while in rank four we have the tensors

$$\Delta = \lambda_1 \delta_{\alpha\beta} \delta_{\gamma\delta} + \lambda_2 \delta_{\alpha\gamma} \delta_{\beta\delta} + \lambda_3 \delta_{\alpha\delta} \delta_{\beta\gamma} \tag{A.17}$$

with arbitrary coefficients λ_i. Below we give the proof that we have found *all* the isotropic tensors in rank two and in rank four we deal with special symmetric cases. We leave the study of rank zero, one, three and the general rank four case as exercises.

Isotropy means that equation (A.7) with $\mathbf{T'} = \mathbf{T}$ holds for *all* rotations and reflections in $O(D)$. Thus invariance under any specific reflection (including parity changes and coordinate permutation) and rotation is a *necessary* condition. To construct the proof, we first show that the above tensors $\mathbf{I}$ and Δ are indeed isotropic. Then we need only find enough necessary conditions (i.e. enough specific instances of equation (A.7)) to reduce isotropic tensors to $\mathbf{I}$ or Δ. In rank two equation (A.7) reads

$$I'_{\alpha\beta} = R_{\alpha\alpha'} R_{\beta\beta'} \delta_{\alpha'\beta'} \tag{A.18}$$

or

$$\mathbf{I'} = \mathbf{R}\mathbf{R}^T, \tag{A.19}$$

thus from (A.2) $\mathbf{I'} = \mathbf{I}$. In rank four we do the special case $\Delta_{\alpha\beta\gamma\delta} = \delta_{\alpha\beta}\delta_{\gamma\delta}$

$$\Delta'_{\alpha\beta\gamma\delta} = R_{\alpha\alpha'} R_{\beta\beta'} R_{\gamma\gamma'} R_{\delta\delta'} \delta_{\alpha'\beta'} \delta_{\gamma'\delta'}. \tag{A.20}$$

Rearranging the right-hand side we let appear twice the product $\mathbf{R}^T\mathbf{R}$ and by (A.2) we get $\Delta'_{\alpha\beta\gamma\delta} = \delta_{\alpha\beta}\delta_{\gamma\delta}$. The proof for the other cases is identical.

We now turn to the necessary conditions. It is interesting to discuss first invariance by isometries in the octahedral group defined in Section A.1. In rank two invariance of $\mathbf{T}$ by parity transformations immediately yields $T_{12} = -T_{12} = 0$. Applying the coordinate permutation $\mathbf{P}_{12}$, we get $T_{22} = T_{11}$ and more generally from cubic symmetry $T_{\alpha\alpha} = T_{11}$ for all α. This last result could also be obtained from the $2\pi/3$ rotation. From (A.8) and $T_{12} = T_{21} = 0$, we get $T_{11} = \frac{1}{4}T_{11} + \frac{3}{4}T_{22}$ and we have again $T_{11} = T_{22}$.

For tensors of rank four we only need, through the considerations of Section A.2, to consider the case in which we can permute the *first* two indices. The sufficient condition for isotropy is equation (A.17) and invariance by permutation of the first two indices yields $\lambda_2 \delta_{\alpha\gamma}\delta_{\beta\delta} + \lambda_3 \delta_{\alpha\delta}\delta_{\beta\gamma} = \lambda_3 \delta_{\alpha\gamma}\delta_{\beta\delta} + \lambda_2 \delta_{\alpha\delta}\delta_{\beta\gamma}$ and thus $\lambda_2 = \lambda_3$ and the tensor is also invariant under permutaion of the *last* two indices.

For simplicity however we will only consider the case in which the tensor is invariant under permutation of all four indices. Using cubic symmetry we find only four independent terms T_{1111}, T_{1212}, T_{1221}, and T_{1122}. Invariance under permutation of all the indices reduces the four independent terms to two distinct terms T_{1111} and $T_{1122} = T_{1212} = T_{1221}$. Thus tensors with cubic symmetry are reduced to a 2D vector space in which all components are equal to T_{1111} or T_{1122}. To show that no further reductions are possible, consider the tensor

$$T_{\alpha\beta\gamma\delta} = \lambda(\delta_{\alpha\beta}\delta_{\gamma\delta} + \delta_{\alpha\gamma}\delta_{\beta\delta} + \delta_{\alpha\delta}\delta_{\beta\gamma}) + \mu\delta_{\alpha\beta}\delta_{\gamma\delta}\delta_{\alpha\gamma} \qquad (A.21)$$

with arbitrary coefficients λ and μ. We have already shown that the first term is isotropic. The term $\mu\delta_{\alpha\beta}\delta_{\gamma\delta}\delta_{\alpha\gamma}$ is simply the tensor where the components with all indices identical (T_{1111}, T_{2222}, etc...) are all equal to μ and all other components are 0. It is invariant by all the $\mathbf{P}_i$ and $\mathbf{P}_{ij}$ transformations, and thus by the whole octahedral group. These two terms are independent, because they span the entire 2D subspace defined above. Indeed $T_{1111} = 3\lambda + \mu$ and $T_{1122} = \lambda$, which may be inverted to

$$\lambda = T_{1122} \quad \text{and} \quad \mu = T_{1111} - 3T_{1122}. \qquad (A.22)$$

Since the two terms are independent, the space of symmetric tensors is at least of dimension two, but we found above that it is at most of dimension two, so we have characterized all tensors with cubic symmetry in the form (A.21).

Consider now the invariance of a rank four tensor under all isometries. In addition to cubic symmetry we may consider invariance under a $2\pi/3$ rotation. Using equation (A.7) applied to the component T_{1111} we find

$$T_{1111} = \frac{1}{16}T_{1111} + \frac{9}{8}T_{1122} + \frac{9}{16}T_{2222}, \qquad (A.23)$$

and

$$T_{1111} = 3T_{1122}. \qquad (A.24)$$

If we use the characterizations (A.21) of cubic symmetry then (A.22) yields $\mu = 0$. Thus the general form of an isotropic rank four tensor, invariant under permutation of the indices is

$$T_{\alpha\beta\gamma\delta} = \lambda(\delta_{\alpha\beta}\delta_{\gamma\delta} + \delta_{\alpha\gamma}\delta_{\beta\delta} + \delta_{\alpha\delta}\delta_{\beta\gamma}). \qquad (A.25)$$

Notice that the proof may extended without change to any dimension D.

A.4   Symmetries of tensors associated with a lattice

Lattice gas systems keep the symmetry of the underlying lattice. In other words, if a lattice gas is rotated and the lattice gas falls back into place, the new system cannot statistically be distinguished from the former one. This is true of the system generally, with all its possible statistical states. Obviously, a particular state, for instance with a given uniform velocity $\mathbf{u}$, is not isotropic. However, the connection between **u** and other properties of the system, such as the momentum flux tensor, is a *general* property of the system, and is invariant by rotation. Thus if the relation

$$\Pi_{\alpha\beta} = T_{\alpha\beta\gamma\delta}u_\gamma u_\delta \qquad (A.26)$$

holds with some constant tensor $T_{\alpha\beta\gamma\delta}$, this tensor generally characterizes the lattice and should be invariant by all lattice isometries. It is therefore of crucial importance to know whether the lattice isometries contain sufficient symmetries (i.e. enough necessary conditions in the sense of the above section) such that **T** is isotropic. Below we show that it is indeed the case on the hexagonal and FCHC lattices.[§]

The FCHC lattice has hypercubic symmetry which readily gives isotropic rank two tensors and equation (A.21) for fourth rank tensors. We shall see below it also gives isotropy for rank four. The hexagonal case is a bit trickier. For rank two tensors it is readily treated as in Section A.2 using parity and $2\pi/3$ rotations. We find again that invariant rank two tensors are of the form $\delta_{\alpha\beta}$. For fourth rank tensors we use parity in $\mathcal{R}$ (but not permutations in $\mathcal{T}_\sigma$, which are not allowed), and permutation of the indices to find three independent terms, T_{1111}, T_{2222} and T_{1122}. Invariance by $2\pi/3$ rotations yields equation (A.23) and thus

$$\frac{15}{16}T_{1111} = \frac{9}{8}T_{1122} + \frac{9}{16}T_{2222}, \qquad (A.27)$$

while applying the same invariance to the component T_{2222}

$$\frac{15}{16}T_{2222} = \frac{9}{8}T_{1122} + \frac{9}{16}T_{1111}. \qquad (A.28)$$

Subtracting we find $T_{1111} = T_{2222}$ and we have cubic symmetry. As above, equation (A.23) then yields $T_{1111} = 3T_{1122}$ and thus isotropy.

In the FCHC case we have cubic symmetry and symmetry about the plane defined by $x_1 + x_2 + x_3 + x_4 = 0$, i. e. the transformation

$$\mathbf{S} : x_\alpha \to x'_\alpha = x_\alpha - \frac{1}{2}\sum_\beta x_\beta. \qquad (A.29)$$

We could write the action of **S** directly using (A.7) but there is a shortcut. The scalar obtained by *contracting* **T** with the vector $\mathbf{y} = (2,0,0,0)$ is an object $s = y_\alpha y_\beta y_\gamma y_\delta T_{\alpha\beta\gamma\delta}$ obtained by repeated scalar products. Since s is a scalar it is isotropic. Writing the invariance of s under the transformation **S** we obtain

$$y_\alpha y_\beta y_\gamma y_\delta T_{\alpha\beta\gamma\delta} = y'_\alpha y'_\beta y'_\gamma y'_\delta T_{\alpha\beta\gamma\delta} \qquad (A.30)$$

[§] We have already seen that **T** is not in general isotropic when the lattice has only cubic symmetry.

where $\mathbf{Sy} = \mathbf{y'} = (1, -1, -1, -1)$. Inserting (A.21) we find 16μ on the left-hand side and 4μ on the right-hand side of (A.30). Thus $\mu = 0$ and $\mathbf{T}$ is isotropic.

A.5 Tensors formed with generating vectors

Lattice gas theory involves the rth order tensors

$$C^{(r)}_{\alpha_1 \ldots \alpha_r} = \sum_i c_{i\alpha_1} \ldots c_{i\alpha_r} \tag{A.31}$$

in which all vectors $\mathbf{c}_i$ have the same norm c. Moreover the set $\{\mathbf{c}_i\}$ is invariant under the lattice symmetry group $\mathcal{G}$ and thus any tensors formed from these generating vectors must have the same symmetries! We are now able to determine these tensors to order 3:

$$\sum_i c_{i\alpha} = 0, \tag{A.32}$$

$$\sum_i c_{i\alpha} c_{i\beta} = \frac{bc^2}{D} \delta_{\alpha\beta}, \tag{A.33}$$

$$\sum_i c_{i\alpha} c_{i\beta} c_{i\gamma} = 0. \tag{A.34}$$

Expressions (A.32) and (A.34) are obtained by parity. It suffices to remark that vectors $\mathbf{c}_i$ appear in pairs: for each $\mathbf{c}_i$ there is another opposed vector $\mathbf{c}_j = -\mathbf{c}_i$. To derive expression (A.33) we first remark that the tensor on the left-hand side is invariant by the symmetry group $\mathcal{G}$ of the lattice. Since all the lattices we consider have at least hypercubic symmetry the results of Section A.4 imply that $C^{(2)}$ is proportional to $\mathbf{I}$. The coefficient of proportionality is readily found by evaluation of $C^{(2)}_{\alpha\alpha}$.

For those lattices which yield isotropic fourth order tensors we must also have

$$C^{(4)}_{\alpha\beta\gamma\delta} = \lambda(\delta_{\alpha\beta}\delta_{\gamma\delta} + \delta_{\alpha\gamma}\delta_{\beta\delta} + \delta_{\alpha\delta}\delta_{\beta\gamma}). \tag{A.35}$$

The proportionality constant λ is determined by summing (A.31) over paired indices to yield (where in the line just below, for clarity, we abandon Einstein's notation)

$$\sum_{\alpha\beta} C^{(4)}_{\alpha\alpha\beta\beta} = \sum_i \left(\sum_\alpha c_{i\alpha} c_{i\alpha} \right) \left(\sum_\beta c_{i\beta} c_{i\beta} \right) \tag{A.36}$$

$$= \sum_i \mathbf{c}_i^2 \mathbf{c}_i^2 \tag{A.37}$$

$$= bc^4. \tag{A.38}$$

From (A.35) (Einstein's notation is back)

$$C^{(4)}_{\alpha\alpha\beta\beta} = \lambda(\delta_{\alpha\alpha}\delta_{\beta\beta} + 2\delta^2_{\alpha\beta}) \tag{A.39}$$
$$= \lambda D(D+2). \tag{A.40}$$

Thus

$$\sum_i c_{i\alpha}c_{i\beta}c_{i\gamma}c_{i\delta} = \frac{bc^4}{D(D+2)}(\delta_{\alpha\beta}\delta_{\gamma\delta} + \delta_{\alpha\gamma}\delta_{\beta\delta} + \delta_{\alpha\delta}\delta_{\beta\gamma}). \tag{A.41}$$

A.6   Tensor attached to a given lattice vector

It is of interest to determine the general form of a tensor $t_{i\alpha\beta}$ attached to the lattice vector c_i and symmetric in the indices $\alpha\beta$. This extends the results obtained in Section 3.2 for square lattices, and is useful in the derivation of the viscosity tensor in Section 15.6. A tensor attached to a vector c_i is invariant by lattice symmetries in the polytope (or lattice) symmetry group $\mathcal{G}$ that leave c_i invariant. Its form is of interest to determine the general form of perturbations at first order in the Chapman–Enskog expansion. The form of the tensor is determined by the following theorem.

Theorem A.1 *Let $\mathcal{L}$ be a regular Bravais lattice and let c_i be the vectors joining the nearest neighbors. Let $\mathcal{G}$ be the symmetry group of the regular polytope formed by the c_i. Let $t_{i\alpha\beta}$ be a tensor symmetric in the indices and invariant by all symmetries in $\mathcal{G}$ leaving c_i fixed. Then $t_{i\alpha\beta}$ is of the form*

$$t_{i\alpha\beta} = \lambda c_{i\alpha}c_{i\beta} + \mu\delta_{\alpha\beta}. \tag{A.42}$$

Proof: Let us first recall some useful definitions. In this section we will find it more convenient to discuss symmetry groups of the lattice from the point of view of the D-dimensional polytope formed by the c_i, although the polytope and the lattice have the same symmetry group $\mathcal{G}$, first mentioned in Chapter 5. A polytope is a high-dimensional generalization of polygons or polyhedra. (The general study of their symmetry groups would take another book, but here we shall be content with a few simple facts.) A set of transformations of space R^D which leaves no linear subspace invariant is called an *irreducible family*. To illustrate, consider a group $\{A(s)\}$ of linear operators transforming space R^D, parameterized by an index s. If there is an invariant linear subspace V for all matrices

$\mathbf{A}(s)$ then they may be written in the appropriate basis in block form

$$\mathbf{A}(s) = \begin{pmatrix} \mathbf{A}_1(s) & \mathbf{A}_2(s) \\ 0 & \mathbf{A}_4(s) \end{pmatrix} \tag{A.43}$$

where $\mathbf{A}_i(s)$ is a block matrix. For instance the continuous group of rotations about an axis leaves that axis invariant and is thus reducible.

Lemma A.1 *The symmetry group $\mathcal{G}$ of a regular polytope is an irreducible family.*

A proof of the lemma for all regular polytopes requires a systematic study of D-dimensional polytopes but here we shall only need it for the polytopes in Table 5.1. Then it is easy to find enough symmetry operations to ensure irreducibility. For instance invariance under the group of all coordinate reflections $\mathcal{R}$ is enough to ensure irreducibility.

Lemma A.2 (Schurr) *If a transformation commutes with all transformations in an irreducible family it is proportional to the identity.*

A proof of this famous lemma of the representation theory of groups may be found in Refs. [A.3,A.4].

We consider the decomposition of space into $\mathbf{c}_i$ and an orthogonal hyperplane Π_i. Before finally proving the theorem we need to prove that $t_{i\alpha\beta}$ has the block structure[¶]

$$\mathbf{T}_i = \begin{pmatrix} \lambda & 0 \\ 0 & \mathbf{B} \end{pmatrix} \tag{A.44}$$

where λ is a scalar coefficient and $\mathbf{B}$ is the matrix of a transformation acting on Π_i.

To prove (A.44), let us show first that $\mathbf{T}_i$ leaves $\mathbf{c}_i$ invariant. We reason by the absurd and consider a decomposition of the vector $\mathbf{T}_i\mathbf{c}_i$ into a component parallel to $\mathbf{c}_i$ and another non-zero vector $\mathbf{v}_i$ in Π_i:

$$\mathbf{T}_i\mathbf{c}_i = \lambda\mathbf{c}_i + \mathbf{v}_i. \tag{A.45}$$

The transformations leaving $\mathbf{c}_i$ invariant and the other vectors globally invariant form a subgroup $\mathcal{G}_i$ of the polytope symmetry group $\mathcal{G}$. $\mathcal{G}_i$ is the symmetry group of the *vertex figure P_i* (the vertex figure is the $D-1$ dimensional polytope formed by the midpoints of the edges attached to $\mathbf{c}_i$). Let $\mathbf{R}$ be a transformation in $\mathcal{G}_i$. The hypothesis in the theorem is

---

[¶] We can identify tensors and linear operators because we consider orthogonal transformations only.

that $t_{i\alpha\beta}$ is invariant by symmetries in $\mathcal{G}_i$. This means that $\mathbf{T}_i$ commutes with $\mathbf{R}$ and thus from (A.45)

$$\mathbf{R}^{-1}\mathbf{T}_i\mathbf{R}\mathbf{c}_i = \lambda\mathbf{c}_i + \mathbf{v}_i. \tag{A.46}$$

The transformation $\mathbf{R}$ leaves $\mathbf{c}_i$ invariant; thus

$$\mathbf{T}_i\mathbf{c}_i = \lambda\mathbf{c}_i + \mathbf{R}\mathbf{v}_i \tag{A.47}$$

and we obtain $\mathbf{R}\mathbf{v}_i = \mathbf{v}_i$ for all $\mathbf{R}$ in $\mathcal{G}_i$. The vector $\mathbf{v}_i$ generates a linear subspace V invariant by all transformations in $\mathcal{G}_i$. Thus $\mathcal{G}_i$ is reducible. But $\mathcal{G}_i$ is the symmetry group of the vertex figure P_i. The vertex figure of a D dimensional regular polytope is a $D-1$ dimensional regular polytope and is thus irreducible from Lemma A.1. We arrive at a contradiction.

Thus $\mathbf{T}_i$ leaves $\mathbf{c}_i$ invariant and may be written in block form

$$\mathbf{T}_i = \begin{pmatrix} \lambda & \mathbf{A}_1 \\ 0 & \mathbf{B} \end{pmatrix} \tag{A.48}$$

for some matrices $\mathbf{A}_1$ and $\mathbf{B}$. Since $\mathbf{T}_i$ is symmetric, $\mathbf{A}_1 = 0$ and $\mathbf{T}_i$ is of the form (A.44). Thus $\mathbf{T}_i$ leaves Π_i globally invariant.

We may now prove the theorem. Since $\mathbf{T}_i$ commutes with all transformations in $\mathcal{G}_i$, so does $\mathbf{B}$. Then by Schurr's lemma $\mathbf{B} = \mu\mathbf{I}$ for some scalar μ. Then from (A.44) we find the expression in the theorem. ∎

A.7   Notes

A useful discussion of tensors may be found in:

A.1 Aris, R. (1962). *Vectors, Tensors, and the Basic Equations of Fluid Mechanics* (Prentice-Hall).

An introduction to symmetries of polygons and polyhedra may be found in:

A.2 *Handbook of Applicable Mathematics, volume V: Combinatorics and Geometry, part B.* W. Ledermand and S. Vajda, editors (Wiley, New York).

Two sources for Schurr's lemma are

A.3 Boerner, H. (1955). *Group Representations* (Springer, Berlin).

A.4 Fulton, W. and Harris, J. (1991). *Representation Theory—A First Course, Graduate Texts in Mathematics, Number 129* (Springer, Berlin).

Appendix B
Polytopes and their symmetry group

The purpose of this appendix is to discuss the structure of the symmetry group of the FCHC lattice. There is a useful decomposition of this symmetry group that is often used in constructing three-dimensional lattice-gas algorithms. However the construction itself is never discussed in the literature. We shall first define polytopes and the Schläfi symbol, and then construct the symmetry group step by step, starting with the triangular and cubic groups and finishing with the hypercubic and FCHC symmetry groups.

B.1 Polytopes and the Schläfi symbol

We start with some definitions. Polytopes are ensembles of vertices. One defines *edges* joining nearest-neighbor vertices, *faces* which are the polygons bounded by the edges, and three-dimensional *cells* bounded by edges and faces.

Let A be a vertex of the polytope Π. The set of points nearest to A on the polytope is the star-polytope Π_A^*. For instance the star polytope of the cube is the triangle. The *Schläfi symbol* $\{p, q, r, \ldots, z\}$ of a regular polytope Π is then formed by the number of edges p of a face and the Schläfi symbol $\{q, r, \ldots, z\}$ of Π^*. This notation is easily applied to polygons and polyhedra. The symbol of a polygon with n vertices is $\{n\}$. The five regular polyhedra or *Platonic solids* are the tetrahedron $\{3,3\}$, octahedron $\{3,4\}$, cube $\{4,3\}$, dodecahedron $\{5,3\}$, and icosahedron $\{3,5\}$. In that case the *Schläfi symbol* $\{p, q\}$ indicates the number p of edges around each face and the number q of edges attached to each vertex.

In four dimensions, we may define the hypercube P_{16} as the 16 points

$$\mathbf{e}_i = (\pm 1, \pm 1, \pm 1, \pm 1). \tag{B.1}$$

Define the point $O = (1, 1, 1, 1)$. The four nearest neighbors of O are

Table B.1.   The nearest neighbors of $(1,1,0,0)$ in P_{24}

(1,0,1,0)
(1,0,-1,0)
(1,0,0,1)
(1,0,0,-1)
(0,1,1,0)
(0,1,-1,0)
(0,1,0,1)
(0,1,0,-1)

the point $(-1, 1, 1, 1)$ and three other points obtained by permutations. Thus the star polytope forms a tetrahedron $\{3, 3\}$. A cell of P_{16} is the cube obtained by intersecting P_{16} with the hyperplane $x_1 = 1$, its elements being $\{(1, \pm1, \pm1, \pm1)\}$. The faces are squares and thus the Schläfi symbol of the hypercube is $\{4, 3, 3\}$. A dual polytope is obtained by taking the nodes on the centers of these faces. It has eight elements of the form

$$\mathrm{perm}(\pm1, 0, 0, 0)$$

The star polytope is an octahedron $\{3, 4\}$. With the same method as above we find that the faces are triangles and thus the Schläfi symbol is $\{3, 3, 4\}$.

Consider now the FCHC polytope. P_{24} is made of the vertices given in equation (5.4). Its star polytope is a cube: an example is obtained by listing all the nearest neighbors of $A = (1, 1, 0, 0)$ in Table B.1. We now try to find the cells of P_{24}. Three of the nearest neighbors of A are together with A in the hyperplane $\sum_i x_i = 2$. This hyperplane contains a cell of P_{24}. The elements of this cell are all the elements of P_{24} verifying $\sum_i x_i = 2$ and they are given in Table B.2. It is easy to check that these eight vertices form an octahedron. It is found that P_{24} has 24 such cells. Finally the faces of P_{24} are also the faces of the octahedron: they are thus triangular and the symbol of P_{24} is $\{3,4,3\}$.

B.2   The orbit-stabiliser theorem

Here we prove a useful theorem, allowing us to decompose the group structure. We first give some definitions. Let $\mathcal{G}$ be the symmetry group of a polytope P. The *stabiliser* of a vertex A is the subgroup $\mathcal{G}_A^*$ of $\mathcal{G}$

Table B.2. A cell of P_{24} is formed by all the points intersecting the hyperplane $\sum_i x_i = 2$. It is an octahedron with points listed below.

$$
\begin{array}{c}
(1,1,0,0) \\
(1,0,1,0) \\
(0,1,1,0) \\
(1,0,0,1) \\
(0,0,1,1) \\
(0,1,0,1)
\end{array}
$$

leaving A invariant. The *orbit* of A is the set of vertices of the form gA where g belongs to $\mathcal{G}$. For a regular polytope the orbit of A is P.

Theorem B.1 *Let A be a fixed element of a polytope P. Let $\mathcal{G}_A^*$ be its stabiliser. For any point B let $\hat{\mathcal{G}}_{AB}$ be the set of isometries that send A on B. For each B we pick a specific isometry $\hat{g}_{AB}$ in $\hat{\mathcal{G}}_{AB}$. Let $\hat{\mathcal{G}}_A$ be the set of isometries $\hat{g}_{AB}$. Then every element of the symmetry group $\mathcal{G}$ of P may be written in a unique manner as $g = \hat{g}g'_A$ where $\hat{g} \in \hat{\mathcal{G}}_A$ and $g'_A \in \mathcal{G}_A^*$. Similarly each element g of $\mathcal{G}$ may be written uniquely as $g = g'_A\hat{g}$ where $\hat{g} \in \hat{\mathcal{G}}_A$ and $g'_A \in \mathcal{G}_A^*$.*

We prove the theorem as follows. Pick a point A. We let $B = g(A)$. Then $\hat{g}$ must be the unique isometry $\hat{g} = \hat{g}_{AB}$ in $\hat{\mathcal{G}}_A$ sending A on B. Also we must have $g'_A = \hat{g}_{AB}^{-1}g$. It is clear that $g'_A \in \mathcal{G}_A^*$. Together with a similar proof for the factorization $g = g'_A\hat{g}$, this proves the theorem.

A corollary is the orbit-stabiliser theorem for the regular polytope:

Theorem B.2 (Orbit-stabiliser theorem) *Let $\mathcal{G}$ be the group of isometries of a polytope, $\mathcal{G}_A^*$ the stabiliser of A and M the orbit of A through $\mathcal{G}$. Then the order of $\mathcal{G}$ is the product*

$$|\mathcal{G}| = |M||\mathcal{G}_A^*|. \tag{B.2}$$

To prove the theorem, we notice that the orbit of A in a regular polytope is the whole polytope. Thus $|M|$ is the number of vertices. But it is also the number of elements in the set $\hat{\mathcal{G}}_A$ defined above. The orbit-stabiliser theorem is then deduced from Theorem B.1.

Using the orbit-stabiliser theorem one may find the orders given in Table 5.1. For instance, the order of the group of $\{3,4,3\}$ is $24 \times 8 \times 3 \times 2 = 1152$. Theorem B.1, on the other hand, may be used to construct the unique factorizations we need to generate isometries in simulations.

B.3   The structure of the {3,4,3} polytope and its symmetry group

To analyze three-dimensional models we need to find the symmetry groups of the three-dimensional and four-dimensional lattices used to build the models. The C, CC and FCC lattices possess the octahedral (cubic) symmetry group. The FCHC lattice has the symmetry group of $\{3, 4, 3\}$. An interesting way to look at $\{3,4,3\}$ is to group the 24 vertices in the following fashion:

$$
\begin{array}{ccc}
A & B & C \\
(1, 0, 0, 1) & (0, 1, 0, 1) & (0, 0, 1, 1) \\
(1, 0, 0, -1) & (0, 1, 0, -1) & (0, 0, 1, -1) \\
(-1, 0, 0, 1) & (0, -1, 0, 1) & (0, 0, -1, 1) \\
(-1, 0, 0, -1) & (0, -1, 0, -1) & (0, 0, -1, -1) \\
(0, 1, 1, 0) & (1, 0, 1, 0) & (1, 1, 0, 0) \\
(0, 1, -1, 0) & (-1, 0, 1, 0) & (1, -1, 0, 0) \\
(0, -1, 1, 0) & (1, 0, -1, 0) & (-1, 1, 0, 0) \\
(0, -1, -1, 0) & (-1, 0, -1, 0) & (-1, -1, 0, 0)
\end{array}
\tag{B.3}
$$

This partitioning of the lattice vectors into three subsets of eight may be motivated by an analysis of the angles between vertex vectors. Indeed, note that if θ_{ij} is the angle between vectors c_i and c_j, the scalar product $\cos\theta_{ij}$ takes values in $\{-2, -1, 0, 1, 2\}$. The $+2$ value is realized only if $i = j$, while -2 implies $c_i = -c_j$. The zero value, corresponding to $\theta_{ij} = \pi/2$ is realized for vectors in the same subgroup that are not equal or opposite. Finally the -1 and 1 values are realized for vectors in different subgroups. While this remark may seem anecdotal, it serves several purposes: it shows that the distinction between A, B and C is intrinsic (they are just the groups of vectors at an angle of zero or π with respect to each other), and not related to a specific basis of four-dimensional space. It also shows that A, B and C, are symmetric to each other. Another useful step to understand this partitioning, and the FCHC symmetry, is to use the isometry

$$
\Lambda = \frac{1}{\sqrt{2}}
\begin{pmatrix}
+1 & +1 & 0 & 0 \\
+1 & -1 & 0 & 0 \\
0 & 0 & +1 & +1 \\
0 & 0 & +1 & -1
\end{pmatrix},
\tag{B.4}
$$

which when applied to each of the 24 lattice vectors in equation (B.3), yields the set of points:*

$$
\begin{array}{ccc}
A & B & C \\
(\,+1,+1,+1,-1\,) & (\,+1,-1,+1,-1\,) & (\ \ 0,\ \ 0,+2,\ \ 0\,) \\
(\,+1,+1,-1,+1\,) & (\,+1,-1,-1,+1\,) & (\ \ 0,\ \ 0,\ \ 0,+2\,) \\
(\,-1,-1,+1,-1\,) & (\,-1,+1,+1,-1\,) & (\ \ 0,\ \ 0,\ \ 0,-2\,) \\
(\,-1,-1,-1,+1\,) & (\,-1,+1,-1,+1\,) & (\ \ 0,\ \ 0,-2,\ \ 0\,) \\
(\,+1,-1,+1,+1\,) & (\,+1,+1,+1,+1\,) & (\,+2,\ \ 0,\ \ 0,\ \ 0\,) \\
(\,+1,-1,-1,-1\,) & (\,-1,-1,+1,+1\,) & (\ \ 0,+2,\ \ 0,\ \ 0\,) \\
(\,-1,+1,+1,+1\,) & (\,+1,+1,-1,-1\,) & (\ \ 0,-2,\ \ 0,\ \ 0\,) \\
(\,-1,+1,-1,-1\,) & (\,-1,-1,-1,-1\,) & (\,-2,\ \ 0,\ \ 0,\ \ 0\,)
\end{array}
. \quad (\text{B.5})
$$

It then appears that $A \cup B$ forms a four-dimensional hypercube, noted $\{4,3,3\}$ in Schläfi notation. Subset C forms the dual polytope $\{3,3,4\}$. By symmetry, all three subsets are $\{3,3,4\}$ polytopes, and any grouping of two is a hypercube $\{4,3,3\}$.

A number of special symmetry operators, which form a set of generators of the symmetry group, are useful to define. Reflections about the hyperplane $x_i = 0$ will be noted $\mathbf{P}_i$ as in (A.4). Reflections about the hyperplanes $x_i - x_j = 0$ exchange the i and j coordinates. They will be noted $\mathbf{P}_{ij}$ as in (A.6). Finally we will note $\mathbf{S}_1$ and $\mathbf{S}_2$ isometries that map A on B or C. We may now state the most useful result of our discussion of symmetry: *Any isometry of $\mathcal{G}$ may be written* uniquely *as the product of elementary reflections and $\mathbf{S}$-isometries* [5.8]:

$$
\begin{pmatrix} \mathbf{I} \\ \mathbf{S}_1 \\ \mathbf{S}_2 \end{pmatrix}
\begin{pmatrix} \mathbf{I} \\ \mathbf{P}_1 \end{pmatrix}
\begin{pmatrix} \mathbf{I} \\ \mathbf{P}_2 \end{pmatrix}
\begin{pmatrix} \mathbf{I} \\ \mathbf{P}_3 \end{pmatrix}
\begin{pmatrix} \mathbf{I} \\ \mathbf{P}_4 \end{pmatrix}
\begin{pmatrix} \mathbf{I} \\ \mathbf{P}_{12} \end{pmatrix}
\begin{pmatrix} \mathbf{I} \\ \mathbf{P}_{13} \\ \mathbf{P}_{23} \end{pmatrix}
\begin{pmatrix} \mathbf{I} \\ \mathbf{P}_{14} \\ \mathbf{P}_{24} \\ \mathbf{P}_{34} \end{pmatrix} . \quad (\text{B.6})
$$

In the above expression, it is understood that one reflection is chosen in each factor.

Proof: We first consider the hypercube $P_{16} = \{4,3,3\}$ defined above. Using the representation (B.1), we attempt to factorize its group $\mathcal{G}_{16}$. As a reference point, pick $A = (1,1,1,1)$. Consider the group $\mathcal{T}_\sigma$ made of isometries that permute the components of a vector $\mathbf{x}$. It leaves A invariant and is thus included in the stabiliser of A. Moreover it is easy

---

* For brevity, we have omitted the normalization factor $1/\sqrt{2}$ in front of each of these vectors.

to check that $\mathcal{T}_\sigma$ has the decomposition

$$\begin{pmatrix} \mathbf{I} \\ \mathbf{P}_{12} \end{pmatrix} \begin{pmatrix} \mathbf{I} \\ \mathbf{P}_{13} \\ \mathbf{P}_{23} \end{pmatrix} \begin{pmatrix} \mathbf{I} \\ \mathbf{P}_{14} \\ \mathbf{P}_{24} \\ \mathbf{P}_{34} \end{pmatrix}, \tag{B.7}$$

and thus has 24 elements. But from the orbit-stabiliser theorem the number of elements of the stabiliser of A is also $|\mathcal{G}_A| = |\mathcal{G}_{16}|/16 = 24$. Thus $\mathcal{T}_\sigma$ is the stabiliser of A. Consider now the group $\mathcal{R}$ generated by the reflections $\mathbf{P}_i$. It has 16 elements and the orbit of A by this group is $\{4,3,3\}$. Thus $\mathcal{R}$ is of the form required by Theorem B.1. It is easy to find that any element of $\mathcal{R}$ may be written as

$$\begin{pmatrix} \mathbf{I} \\ \mathbf{P}_1 \end{pmatrix} \begin{pmatrix} \mathbf{I} \\ \mathbf{P}_2 \end{pmatrix} \begin{pmatrix} \mathbf{I} \\ \mathbf{P}_3 \end{pmatrix} \begin{pmatrix} \mathbf{I} \\ \mathbf{P}_4 \end{pmatrix}. \tag{B.8}$$

Using Theorem B.1 we have thus completed the factorization of the symmetry group $\mathcal{G}_{16}$ of $\{4,3,3\}$, which by symmetry is also the symmetry group of $\{3,3,4\}$. Any element of $\mathcal{G}_{16}$ thus leaves invariant the subsets $A \cup B$ or C defined in equation (B.3). We have altogether $16 \times 24 = 384$ isometries in $\mathcal{G}_{16}$. Factoring by the group of three isometries $(\mathbf{I}, \mathbf{S}_1, \mathbf{S}_2)$ that permute the subsets A, B, or C we obtain the $1152 = 48 \times 24$ isometries of $\{3,4,3\}$, in the form (B.6). ∎

B.4   Notes

A fascinating account of polytopes may be found in:

B.1 Coxeter, H. S. M. (1977). *Regular Polytopes* (Dover, New York).

Our description of the three-block structure of FCHC is inspired by Ref. [5.8]. A factorization similar to the above was first given without proof in:

B.2 Hénon, M. (1987) Isometric collision rules for the four-dimensional FCHC lattice gas. *Complex Systems* 1, 475–494.

There are similar factorizations in Refs. [5.7] and [B.2], which are based on the same "hypercubic" symmetry group $\mathcal{G}_{16}$.

Appendix C
Classical compressible
flow modeling

C.1   Non-dissipative, inviscid, compressible flow

We consider a simple fluid such as air or water. This fluid is described by a number of thermodynamical fields, such as pressure, density, etc., as well as a velocity field $\mathbf{u}$. A variable is called *specific* when it gives a quantity per unit mass. For instance, let E be the internal energy of a finite volume V of fluid. Let $M = \rho V$ be the mass of this volume of fluid. Then $e = E/M$ is the specific internal energy. Table C.1 lists all the thermodynamic variables used in this appendix.

In addition to thermodynamic variables there are variables describing the external actions on the fluid. The effect of gravity, for instance may be represented by the acceleration $\mathbf{f} = \mathbf{g}$. The heating rate per unit mass q represents sources of heat, for instance from radiation. There may be heat and momentum exchanges inside the fluid, by heat conduction or viscous forces. These are not taken into account in the non-dissipative description.

For a one-component gas, there are only two independent thermodynamic variables. As a special choice, one may choose ρ, T as independent variables and express all quantities such as p, e, h, etc., in terms of ρ and T, but other choices are possible as well. These relations are linked through equations of state. For instance $p = p(\rho, T)$ is the standard form for an equation of state. In a perfect gas

$$p = \rho R T \qquad (C.1)$$

where R is a constant, which is a characteristic of the gas under study.

Table C.1.   Notations

γ	C_p/C_v
κ	heat conductivity
λ	second viscosity coefficient
μ	first viscosity coefficient
ρ	density
C_p	specific heat at constant pressure
C_v	specific heat at constant volume
e	specific internal energy
$\mathbf{f}$	force per unit mass of fluid
h	specific enthalpy $(=e+p/\rho)$
k	specific mechanical energy $(=e+1/2u^2)$
p	pressure
q	specific heating rate
s	specific entropy
T	temperature
u^2	velocity squared $u_\alpha u_\alpha$

The balance of momentum given in Chapter 2, as well as the conservation of mass and mechanical energy, leads to the system of equations

$$\partial_t \rho + \partial_\alpha(\rho u_\alpha) = 0 \tag{C.2}$$

$$\partial_t(\rho u_\alpha) + \partial_\beta(\rho u_\alpha u_\beta) = -\partial_\alpha p + \rho f_\alpha \tag{C.3}$$

$$\partial_t(\rho e + \rho\frac{u^2}{2}) + \partial_\alpha[(\rho e + \rho\frac{u^2}{2})u_\alpha] = -\partial_\alpha(p u_\alpha) + \rho q + \rho f_\alpha u_\alpha \tag{C.4}$$

These equations are closed by an equation of state of the form $p = p(\rho, e)$. They are identical to equation (2.17) but for the addition of the energy equation (C.4) and the external force term ρf_α. The assumptions leading to equation (C.4) are

- Absence of viscous dissipation

- Absence of heat conduction

- Equivalence of the thermodynamic pressure with the pressure appearing in the dynamical equations

although we hasten to point out that the derivation is complex. The reader is referred to standard references on this topic, such as [3.6].

In the interesting special case where $\mathbf{f} = 0$ and $q = 0$ the energy equation expresses the conservation of energy. More generally, in that case the equations are examples of a system of *conservations laws*. It is possible to obtain from the basic equations the alternative energy equation

$$\frac{de}{dt} + \frac{p}{\rho} \operatorname{div} \mathbf{u} = q + f_\alpha u_\alpha. \tag{C.5}$$

This equation is obtained through some manipulation, after multiplying equation (C.3) by u_α and using the continuity equation (C.2). A case of particular interest is a flow with $q = 0$. It is then useful to recall the thermodynamic identity that defines the entropy. In a quasi-reversible process

$$de = T\, ds + \delta w, \tag{C.6}$$

where $\delta w = -p\,d(1/\rho) + f_\alpha u_\alpha dt$ is the total work on the gas particle done by pressure and external forces. If we identify the variation de with the energy variation of a particle moving with the flow, and after some manipulation involving (C.2), we obtain

$$\frac{ds}{dt} = 0. \tag{C.7}$$

Equation (C.7) expresses then that if the entropy is initially uniform, with $s = s_0$ and s_0 a constant it will remain so. When the entropy is uniform, the fluid is fully described by equations (C.2) and (C.3) with $p = p(\rho, s_0)$. An often used simple form for $p(\rho, s_0)$ is that of a *polytropic gas*. In this case

$$p = A\rho^\gamma, \tag{C.8}$$

where A is a constant depending on the entropy s_0. Analysis of sound wave propagation in equations (C.2), (C.3) and (C.7) may be done in the manner of Appendix D. One finds that the speed of sound is in the general case is

$$c_s^2 = \left(\frac{\partial p}{\partial \rho}\right)_s \tag{C.9}$$

where the subscript s indicated that the derivative is taken at constant entropy s. The particular case of a polytropic gas leads to

$$c_s^2 = \gamma A \rho^{\gamma-1}. \tag{C.10}$$

If the polytropic gas is also a perfect gas one obtains the celebrated formula

$$c_s = \sqrt{\gamma R T}. \tag{C.11}$$

For the polytropic gas with no external heating or forces, equations (C.3) and (C.4) reduce to the single equation

$$\partial_t(\rho u_\alpha) + \partial_\beta(\rho u_\alpha u_\beta) = -\gamma A \rho^{\gamma-1}\partial_\alpha \rho + \rho f_\alpha. \tag{C.12}$$

This equation may be modeled with the lattice-Boltzmann method (see Exercise 6.3).

C.2 Compressible viscous flow in three dimensions of space

We have discussed the Navier-Stokes equation in Chapter 2. For reasons that will soon be clear, it is preferable to discuss the three-dimensional case first. It is useful to rewrite equation (2.30) in the form

$$\partial_t(\rho u_\alpha) + \partial_\beta \rho u_\alpha u_\beta = -\partial_\alpha p - \partial_\beta \Pi_{\alpha\beta}^{visc} + \rho f_\alpha. \tag{C.13}$$

Here the viscous momentum flux may be written as

$$\Pi_{\alpha\beta}^{visc} = -\mu(\partial_\beta u_\alpha + \partial_\alpha u_\beta - \frac{2}{3}\delta_{\alpha\beta}\text{div}\mathbf{u}) - \lambda\delta_{\alpha\beta}\text{div}\mathbf{u} \tag{C.14}$$

where μ is the same coefficient as in equation (2.30), and $\lambda = \xi + \mu/3$. For a perfect gas, classical kinetic theory shows that $\lambda = 0$. Similarly, λ vanishes for single velocity lattice gases in 3D.

Thermal conduction effects appear in the energy equation. We assume Fourier law in the form

$$\mathbf{J} = -\kappa\nabla T, \tag{C.15}$$

where $\mathbf{J}$ is the heat flux and κ the heat conductivity (which is taken to be a constant for simplicity). The energy equation is then modified by the presence of viscous forces and heat conduction, and becomes

$$\partial_t(\rho e + \rho\frac{u^2}{2}) + \partial_\alpha[(\rho e + \rho\frac{u^2}{2})u_\alpha] = -\partial_\alpha(p u_\alpha) + \rho q + \rho f_\alpha u_\alpha$$
$$-\partial_\alpha(\Pi_{\alpha\beta}^{visc}u_\beta) + \kappa\nabla^2 T. \tag{C.16}$$

Using (C.2), (C.13) and (C.16) one obtains an equation for the internal energy in the dissipative case

$$\rho\frac{de}{dt} + p\,\text{div}\mathbf{u} = \rho q + \rho f_\alpha u_\alpha - \Pi_{\alpha\beta}^{visc}\partial_\alpha u_\beta + \kappa\nabla^2 T. \tag{C.17}$$

From (C.17) it is possible to obtain the rate of entropy production using again the definition of entropy (C.6):

$$\rho T\frac{ds}{dt} = -\Pi_{\alpha\beta}^{visc}\partial_\alpha u_\beta + \kappa\nabla^2 T + \rho q. \tag{C.18}$$

Using (C.14) equation (C.18) may be rewritten as

$$\rho T \frac{ds}{dt} = \frac{\mu}{2} \left(\partial_\alpha u_\beta + \partial_\beta u_\alpha - \frac{2}{3} \mathrm{div} \mathbf{u} \delta_{\alpha\beta} \right)^2 + \lambda (\mathrm{div} \mathbf{u})^2 + \kappa \nabla^2 T + \rho q. \quad (C.19)$$

This equation gives the entropy production. From the second principle of thermodynamics, entropy production must be non-negative and hence μ, λ and κ are non-negative.

C.3 Generalization to D dimensions of space

For reasons that will appear later, in the D-dimensional case, the viscous momentum flux must be written as

$$\Pi^{visc}_{\alpha\beta} = -\mu \left(\partial_\beta u_\alpha + \partial_\alpha u_\beta - 2 \frac{D-2}{D} \mathrm{div} \mathbf{u} \delta_{\alpha\beta} \right) - \lambda_D \mathrm{div} \mathbf{u} \delta_{\alpha\beta} \quad (C.20)$$

where μ is the same coefficient as in equation (2.30), and $\lambda_D = \xi + \mu/D$. We have added the subscript D to emphasize the change in the definition of λ. This change stems from the transformation of the coefficient $-2/3$ in (C.14) to $-2(D-2)/D$, and is motivated by the fact that it allows (C.19) to still hold. Without this change, it would be impossible to put the entropy production in the form of a sum of squares as in (C.19), and the positivity of the coefficients would not imply anymore the second principle.

For a single velocity gas in D dimensions, we find again that λ_D vanishes. This may be obtained from classical kinetic theory or from our detailed calculations for the lattice gas. There is however a subtlety: for λ_D to vanish, D must be taken as the dimension of the space of discrete velocities (for instance four in the FCHC case) and not the dimension of the flow (which is three for FCHC projected to 3D). Thus if λ_4 vanishes with the four-dimensional definition for FCHC, the three-dimensional λ does not vanish and is $\lambda = \mu \frac{D-3}{3D}$. Thus care must be exercised when it is stated that the second viscosity coefficient vanishes for single-velocity lattice gases.

Appendix D
Incompressible limit

We describe various asymptotic limits relevant to both the classical equations of Navier-Stokes and Euler and to the lattice gas versions of those equations.

D.1   Space, time and velocity scales

We will consider the mass conservation and Euler equations of the lattice gas in the form derived in Chapter 2 or Appendix C. The equations are

$$\partial_t \rho + \partial_\alpha (\rho u_\alpha) = 0 \tag{D.1}$$

$$\partial_t (\rho u_\alpha) + \partial_\beta (\rho g(\rho) u_\alpha u_\beta) = -\partial_\alpha p \tag{D.2}$$

$$p = p(\rho, u^2). \tag{D.3}$$

The equation for the pressure is either the polytropic one (C.8) or the lattice-gas equation (2.54). We thus deal with the lattice-gas case in general, and with the real case for $g(\rho) = 1$. To investigate the asymptotic limit of a partial differential equation we need to fix the time, space, density and velocity scales. These may be fixed in a number of ways: by initial conditions, by boundary conditions (for instance the spatial scale may be related to the size of obstacles in the flow), by the external forces or by consistency requirements in the equations. To avoid complex discussions, scales are fixed *a priori* in asymptotic analysis, in view of the type (or "branch") of solutions that the physicist wants to investigate. As the solution is constructed, the consistency of these scales is verified.

When a scale is fixed, for instance for length, one must keep in mind that a derivative amounts to a division by that scale of length, and thus we shall write $\partial_x \simeq 1/L$. We will avoid too much formality and simply recall that, in this appendix, our goal is principally to recall the results rather than give a full presentation of them.

276

D.2 Sound waves

We consider a space scale L, a time scale T and a ratio of scales $L/T = c_s$, where $c_s = (\partial p/\partial \rho)^{1/2}$ is, as usual, the speed of sound. A small perturbation of amplitude ϵ from the rest state of constant density ρ_0 is assumed. It has the form

$$\rho(\mathbf{x}, t) = \rho_0 + \epsilon \rho_1(\mathbf{x}, t) + \epsilon^2 \rho_2(\mathbf{x}, t) + \cdots \tag{D.4}$$
$$\mathbf{u}(\mathbf{x}, t) = \epsilon \mathbf{u}_1(\mathbf{x}, t) + \epsilon^2 \mathbf{u}_2(\mathbf{x}, t) + \cdots \tag{D.5}$$

We fix the velocity scale by $|\mathbf{u}_1| \simeq c_s$ and leave the density perturbation scale undetermined for the moment. Inserting the expansion into (D.1) and (D.2) and neglecting all terms of order at least ϵ^2,

$$\epsilon \partial_t \rho_1 + \epsilon \rho_0 \nabla \cdot \mathbf{u}_1 = \mathcal{O}(\epsilon^2) \tag{D.6}$$

$$\epsilon \partial_t \mathbf{u}_1 + \epsilon \frac{1}{\rho_0} \frac{\partial p}{\partial \rho} \nabla \rho_1 = \mathcal{O}(\epsilon^2). \tag{D.7}$$

We notice that the nonlinear terms, of the form $\mathbf{u} \cdot \nabla \mathbf{u}$ or those in the pressure $p(\rho, u^2)$ have disappeared. The scale of density perturbations is fixed by (D.6) to be

$$\rho_1 \simeq |\mathbf{u}_1| \rho_0 T/L, \tag{D.8}$$

which with our assumptions leads to $\rho_1 \simeq \rho_0$. This should reassure us: the scale for ρ_1 is consistent with an expansion of the kind (D.4). Recall that $\partial p/\partial \rho = c_s^2$ (the derivative is taken at constant entropy). From (D.6) and (D.7) one may derive

$$\partial_{tt}^2 \rho_1 - c_s^2 \nabla^2 \rho_1 = 0 \tag{D.9}$$

which is the equation for sound waves with soundspeed c_s.

D.3 Incompressible limit

Again, we fix a space scale L and a time scale T. However the ratio of scales is $L/T = U$ where U is a chosen velocity scale much smaller than the speed of sound. We choose $U \sim \epsilon c_s$, and again expand about a constant density ρ_0. We shall see that the following expansion is then consistent

$$\rho(\mathbf{x}, t) = \rho_0 + \epsilon^2 \rho_1(\mathbf{x}, t) + \epsilon^4 \rho_2(\mathbf{x}, t) + \cdots \tag{D.10}$$
$$\mathbf{u}(\mathbf{x}, t) = \epsilon \mathbf{u}_1(\mathbf{x}, t) + \epsilon^3 \mathbf{u}_2(\mathbf{x}, t) + \cdots \tag{D.11}$$

Because the time scale T is large with respect to the time period of sound waves with the same space scale L, it is useful to introduce a rescaled variable of time $t_1 = \epsilon t$. Then $\partial_t \simeq \epsilon \partial_{t_1}$, and inserting the expansion into the continuity equation (D.1) we get

$$\epsilon \rho_0 \nabla \cdot \mathbf{u}_1 + \epsilon^3 \partial_{t_1} \rho_1 + \epsilon^3 \mathbf{u}_1 \cdot \nabla \rho_1 = \mathcal{O}(\epsilon^5). \tag{D.12}$$

There is a single term of order ϵ which must hence vanish:

$$\nabla \cdot \mathbf{u}_1 = 0. \tag{D.13}$$

Inserting the expansion into (D.2)

$$\epsilon^2 \partial_{t_1} \mathbf{u}_1 + \epsilon^2 g(\rho_0) \mathbf{u}_1 \cdot \nabla \mathbf{u}_1 = -\epsilon^2 \frac{1}{\rho_0} \nabla p_1 + \mathcal{O}(\epsilon^4) \tag{D.14}$$

where we have also expanded the pressure as

$$p = p_0 + \epsilon^2 p_1 + \epsilon^4 p_2 + \cdots \tag{D.15}$$

where

$$p_1 = c_s^2 \rho_1 + A u_1^2 \tag{D.16}$$

where $A = \partial p / \partial u^2$ as in Chapter 2. If we perform a change of variable writing $\mathbf{v} = g(\rho_0)\mathbf{u}_1$ and $p' = g(\rho_0)p_1$ we obtain the incompressible Euler equations

$$\nabla \cdot \mathbf{v} = 0 \tag{D.17}$$

$$\partial_{t_1} \mathbf{v} + \mathbf{v} \cdot \nabla \mathbf{v} = -\frac{1}{\rho} \nabla p'. \tag{D.18}$$

These are the classical incompressible flow equations. An equation of state is not needed for these equations and the fact that p_1 depends on $\mathbf{u}_1$ is in fact irrelevant. In fact it may be shown that p_1 may be eliminated by the incompressibility condition. Taking the divergence of equation (D.18) we find

$$\partial_\alpha \partial_\beta v_\alpha v_\beta = -\frac{1}{\rho} \nabla^2 p'. \tag{D.19}$$

For reasonable boundary conditions on the velocity we can invert for the pressure and find

$$p' = -\rho (\nabla^2)^{-1} \partial_\alpha \partial_\beta v_\alpha v_\beta. \tag{D.20}$$

where the notation $(\nabla^2)^{-1}$ indicates the result of the inversion of the Laplacian ∇^2 with the appropriate boundary conditions. The Euler equation becomes

$$\partial_{t_1} \mathbf{v} + \mathbf{v} \cdot \nabla \mathbf{v} = \nabla (\nabla^2)^{-1} \partial_\alpha \partial_\beta v_\alpha v_\beta. \tag{D.21}$$

Thus, at least formally, the pressure is entirely eliminated from the equations. How is this compatible with the fact that the pressure depends nonlinearly on $\mathbf{u}_1$? To see the answer, we replace by (D.16) in (D.20)

$$c_s^2 \rho_1 + A \mathbf{u}_1^2 = -\rho(\nabla^2)^{-1} \partial_\alpha \partial_\beta u_{1\alpha} u_{1\beta} \qquad \text{(D.22)}$$

this equations appears to fix ρ_1 as a function of $\mathbf{u}_1$. Since ρ_1 was left free so far there is no contradiction. This reasoning also applies to the real fluid case, where we let $A = 0$. Equation (D.22) also allows us to check that ρ_1 is of order 1 as originally assumed.

D.4 Viscous case

When viscous effects are added, several limits may be considered. We now have two dimensionless numbers $M = |\mathbf{u}|/c_s$ and $R = |\mathbf{u}|L/\nu$. We deal first with the case when R is either small or order 1. There are no particular problems with small R . Then the viscous terms come as a small correction to the expansions of section D.3. When R is of order 1, the viscous terms and time scales are of the same order as those based on the velocity alone and the expansions also proceed as in section D.3.

We consider the case $R \ll 1$. We start with the viscous equation (C.13) and to simplify matters and shorten the equations we consider the case $\lambda = 0$. We use a different scaling, which will turn out to give the incompressible solution which we seek:

$$\rho(\mathbf{x}, t) = \rho_0 + \epsilon^2 R^{-1} \rho_1(\mathbf{x}, t) + \cdots \qquad \text{(D.23)}$$

$$\mathbf{u}(\mathbf{x}, t) = \epsilon \mathbf{u}_1(\mathbf{x}, t) + \epsilon^3 \mathbf{u}_2(\mathbf{x}, t) + \cdots. \qquad \text{(D.24)}$$

This expansion is consistent if $\epsilon^2 \ll R$. Inserting (D.23) and (D.24) into the continuity equation (D.1)

$$\epsilon \rho_0 \nabla \cdot \mathbf{u}_1 + \epsilon^3 R^{-1} \partial_{t_1} \rho_1 + \epsilon^3 R^{-1} \mathbf{u}_1 \cdot \nabla \rho_1 = \mathcal{O}(\epsilon^5 R^{-1}). \qquad \text{(D.25)}$$

We obtain again at first order the incompressibility condition

$$\nabla \cdot \mathbf{u}_1 = 0. \qquad \text{(D.26)}$$

Inserting the expansion into (D.2), using (D.16) and after some rearrangement of the terms

$$\epsilon^2 \partial_{t_1} \mathbf{u}_1 + \epsilon^2 g(\rho_0) \mathbf{u}_1 \cdot \nabla \mathbf{u}_1 + \epsilon^2 \frac{A}{\rho_0} \nabla \mathbf{u}_1^2 = -\epsilon^2 R^{-1} \frac{c_s^2}{\rho_0} \nabla \rho_1 + \epsilon \nu \nabla^2 \mathbf{u}_1 + \mathcal{O}(\epsilon^4)$$

$$\text{(D.27)}$$

On the right-hand side we have $\epsilon^2 R^{-1} c_s^2 \nabla \rho_1 / \rho_0 \simeq \epsilon^2 R^{-1} c_s^2 L^{-1}$ while $\epsilon \nu \nabla^2 \mathbf{u}_1 \simeq \epsilon \nu L^{-2}$. But we also have the scaling $\nu/(Lc_s) \simeq \epsilon/R$. Thus the terms on the right-hand side are both of order $\epsilon^2 R^{-1} c_s^2 L^{-1}$. Since $R \ll 1$ they are much larger than the $\mathcal{O}(\epsilon^2)$ terms on the left-hand side, including the nonlinear term in the pressure. The remaining balance is

$$-\nabla p_1 + \mu \nabla^2 \mathbf{u}_1 = 0. \qquad \text{(D.28)}$$

In other words inertia disappears and pressure forces balance viscous forces. To summarize the conditions for that scaling, we have

$$M^2 \ll R \ll 1. \tag{D.29}$$

Another possible scaling is obtained if we take into account the viscous time scale $T_\nu = L^2/\nu$. This time scale is much shorter than the convective time scale $T_u = L/|\mathbf{u}|$, but may be an important transient phenomenon in the context of lattice gas simulations. Then we introduce the new time variable $t_2 = t/T_\nu$ and the time derivatives are of order $\partial_t = \epsilon R^{-1}\partial_{t_2}$. Then inserting (D.23) and (D.24) into the continuity equation (D.1),

$$\epsilon \rho_0 \nabla \cdot \mathbf{u}_1 + \epsilon^3 R^{-2}\partial_{t_2}\rho_1 + \epsilon^3 R^{-1}\mathbf{u}_1 \cdot \nabla\rho_1 = \mathcal{O}(\epsilon^5 R^{-1}). \tag{D.30}$$

Interestingly, the incompressibility condition (D.26) is recovered only if $\epsilon \ll R$. This is a stronger condition than previously. In the momentum balance equation, we get

$$\epsilon^2 R^{-1}\partial_{t_2}\mathbf{u}_1 + \epsilon^2 g(\rho_0)\mathbf{u}_1 \cdot \nabla\mathbf{u}_1 = -\epsilon^2 R^{-1}\frac{c_s^2}{\rho_0}\nabla\rho_1 + \epsilon\nu\nabla^2\mathbf{u}_1 + \mathcal{O}(\epsilon^4). \tag{D.31}$$

The condition $\epsilon \ll R$ implies that inertial terms drop out again. The remaining balance is

$$\partial_{t_2}\mathbf{u}_1 = -\frac{1}{\rho_0}\nabla p_1 + \nu\nabla^2\mathbf{u}_1, \tag{D.32}$$

or, in other words, the incompressible Navier-Stokes equation without the nonlinear term. This scaling is possible if

$$M \ll R \ll 1. \tag{D.33}$$

The time scales in this case are ordered as

$$T_s \ll T_\nu \ll T_u \tag{D.34}$$

where T_s, T_ν, T_u are the sound, viscous and velocity time scales, all calculated with the same length scale L, which is identical to the scale L_{hydro} of Chapter 2.

We mention at last a difficulty with viscous flow coming from kinetic theory. The Maxwell estimate for the viscosity of gases (see Section 4.2) is $\nu \sim v_{th}\ell_{\text{mfp}}$. Thus $MR^{-1} = \ell_{\text{mfp}}/L_{\text{hydro}}$. Assume that microscopic and macroscopic scales are separated (Section 2.1) $\ell_{\text{mfp}} \ll L_{\text{hydro}}$. Then if R is small or order 1 we are always in the incompressible case. Conversely, a viscous, compressible flow is impossible.

Appendix E
Derivation of the
Gibbs distribution

We consider the configuration of Figure 14.7, with a large system Σ containing $\mathcal{M}$ sites, subdivided into p smaller systems S_i each containing $M = \mathcal{M}/p$ sites. Each small system has configurations s that belong to one of a set of mass-momentum packets, or energy levels $S_{M_s, \mathbf{G}_s}$. Since the small systems' size is finite, there is a finite number ω of mass-momentum packets. We label these packets by a single index $0 \le \sigma < \omega$, and we also label the corresponding values of the invariants in such a way that the mass-momentum packet σ corresponds to the small-system mass $M_s = J_\sigma$ and momentum $\mathbf{G}_s = \mathbf{K}_\sigma$ when $s \in \sigma$. We let W_σ be the number of configurations in packet σ, or in the quantum picture the degeneracy of energy level σ. To a configuration of the whole domain Σ there corresponds a distribution of the energy levels among the subsystems. For instance level 1 may occur 3 times, level 2 may occur 4 times, etc. We let n_σ be the number of times level σ occurs, and we call $(n_\sigma)_\sigma$ the spatial frequency distribution.

For a given spatial frequency distribution $(n_1, \cdots, n_\sigma, \cdots, n_\omega)$ there are a number of arrangements of the levels among the subsystems S_1, $S_2, \cdots, S_p$. The constraints on the spatial frequency distributions are that the mass M_Σ and momentum $\mathbf{G}_\Sigma$ of the large system are fixed and the n_σ should sum to the total number of subsystems p. Thus

$$\sum_\sigma n_\sigma J_\sigma = M_\Sigma \qquad \text{(E.1)}$$

$$\sum_\sigma n_\sigma \mathbf{K}_\sigma = \mathbf{G}_\Sigma \qquad \text{(E.2)}$$

$$\sum_\sigma n_\sigma = p. \qquad \text{(E.3)}$$

Combinatorics tells us that the number of arrangements having spatial

281

frequency distribution $(n_\sigma)_\sigma$ is

$$\frac{p!}{n_1!n_2!\cdots n_\omega!} .$$

Each of the arrangements corresponds to several configurations, because there are several (in fact W_σ) configurations in each packet σ. Thus we have the total number of configurations with a given spatial frequency distribution $(n_\sigma)_\sigma$:

$$W(n_1, n_2, \cdots, n_\omega) = \frac{p!}{n_1!n_2!\cdots n_\omega!} W_1^{n_1} W_2^{n_2} \cdots W_\omega^{n_\omega} . \qquad (E.4)$$

The probability of having a spatial frequency $(n_\sigma)_\sigma$ is thus

$$P_\Sigma(n_1, n_2, \cdots, n_\omega) = \frac{W(n_1, \cdots, n_\omega)}{Z} \qquad (E.5)$$

if (E.3) holds, and $P_\Sigma(n_1, n_2, \cdots, n_\omega) = 0$ if (E.3) is violated.

Now we seek the probability $P_S(s)$ for any of the smaller systems S_i being in configuration s. This probability may only be a function f_P of the energy level σ of the small system, as in equation (14.22). Thus we write

$$P_S(s) = f_P(J_\sigma, K_\sigma). \qquad (E.6)$$

Moreover all the small systems are identical and we may average the n_σ. For this we define the probability distribution of n_σ for a given σ:

$$P_1(n_\sigma = k) = \sum_{n_1, \cdots, n_{\sigma-1}, n_{\sigma+1}, \cdots, n_\omega} P_\Sigma(n_1, \cdots, n_{\sigma-1}, k, n_{\sigma+1}, \cdots, n_\omega).$$

$$(E.7)$$

Define the average number of subsystems whose energy level is σ:

$$\overline{n}_\sigma = \sum_k k P_1(n_\sigma = k). \qquad (E.8)$$

If we know the energy level is σ, the probability of seeing the configuration $s \in \sigma$ is

$$P_S(s) = \frac{\overline{n}_\sigma}{pW_\sigma}. \qquad (E.9)$$

Now we use the central fact in this derivation: since the system is large, the probability distribution $P_\Sigma(n_1, \cdots, n_\omega)$ is strongly peaked around the average spatial frequency $\overline{n}_1, \overline{n}_2, \cdots, \overline{n}_\omega$ [E.1]. Moreover the average frequencies $\overline{n}_\sigma$ are themselves large. Thus to find $\overline{n}_\sigma$ we simply need to find the maximum of P_Σ with respect to the constraints (E.3). It is more convenient to find the maximum of $\ln P_\Sigma$. Using the well-known Stirling approximation for the factorial of a large number, $\log n! \approx n \log n - n$, we find

$$\ln P_\Sigma = \sum_\sigma (-n_\sigma \ln n_\sigma + n_\sigma + n_\sigma \ln W_\sigma) + \text{Const.} \tag{E.10}$$

Using the method of Lagrange multipliers this amounts to the maximization of

$$L = \sum_\sigma [-n_\sigma \ln n_\sigma + n_\sigma(\ln W_\sigma + 1 + J_\sigma h + \mathbf{K}_\sigma \cdot \mathbf{q} + \lambda)] \tag{E.11}$$

where h, $\mathbf{q}$ and λ are Lagrange multipliers associated with the mass, momentum and probability summation constraints (E.3). Differentiating with respect to n_σ we find

$$\frac{\partial L}{\partial n_\sigma} = -\ln n_\sigma + J_\sigma h + \mathbf{K}_\sigma \cdot \mathbf{q} + \lambda + \ln W_\sigma. \tag{E.12}$$

To find the maximum of L we impose the condition $\partial L/\partial n_\sigma = 0$, which yields the result

$$\bar{n}_\sigma = \text{Const} \times W_\sigma \exp(-J_\sigma h - \mathbf{K}_\sigma \cdot \mathbf{q} - \lambda). \tag{E.13}$$

The probability of a configuration s in packet σ is thus

$$P_S(s) = \frac{\exp(-J_\sigma h - \mathbf{K}_\sigma \cdot \mathbf{q} - \lambda)}{Z}. \tag{E.14}$$

The normalization constant Z is found as in Chapter 14 using the fact that $\sum_s P_S(s) = 1$. As expected from equation (E.6), P_S is a function of the invariants only. This yields the Gibbs distribution (14.23).

Note

The derivation in this appendix is patterned after that found on pages 1–12 of:

E.1 Hill, T. R. (1986). *An Introduction to Statistical Thermodynamics* (Dover, New York).

Appendix F
Hydrodynamic response
to forces at fluid interfaces

In this appendix we derive the frequency-domain expression for the hydrodynamic response function that appears in the interface equation of motion (17.43).

We consider an interface $h(x, t)$ between two fluids in 2D, as shown in Figure 17.3. We define the *stream function* Ψ such that the horizontal and vertical components of the fluid velocity, u_x and u_y, are given by

$$u_x = \partial_y \Psi, \qquad u_y = -\partial_x \Psi. \tag{F.1}$$

An arbitrary force F of wavenumber k and frequency ω acting on the fluid at the position of the interface takes the form

$$F(x, t) = F_0\, \mathbf{e}_y\, e^{i(kx - \omega t)} \tag{F.2}$$

where $\mathbf{e}_y$ is the unit vector in the y-direction. Here we have employed a linearized approximation in which the interface is approximated by the line $y = 0$. Thus under the usual assumption that surface tension does not vary in the x-direction, the force F has non-vanishing components only in the y-direction.*

We also have the following boundary conditions at the interface. First, the continuity of velocity requires that

$$[\partial_x \Psi] = [\partial_y \Psi] = 0 \tag{F.3}$$

where the square brackets indicate the jump conditions defined in Section 9.4, where here side 1 is associated with positive y. Then from equations (F.1) and (F.3) we have

$$[\partial_y u_y] = -[\partial_y \partial_x \Psi] = 0. \tag{F.4}$$

---

* In the case of the lattice gas, the fluctuating nature of the dynamics near the interface causes the surface tension to vary randomly in space and time around a mean value. In particular, the surface tension varies along the x-direction, so a small tangential force exists and there could be some corrections to the analysis presented here.

The continuity of tangential stress requires that

$$[S_{xy}] = 0, \tag{F.5}$$

where

$$S_{\alpha\beta} = -p\delta_{\alpha\beta} + \rho\nu(\partial_\alpha u_\beta + \partial_\beta u_\alpha) \tag{F.6}$$

is the stress tensor. Note that the force (F.2) at the interface has the same sign as the jump in pressure. Then from equations (F.4) and (F.6), we have

$$[S_{yy}] = -F_0 e^{i(kx-\omega t)}. \tag{F.7}$$

Also, symmetry with respect to the interface gives

$$\Psi(x, y) = \Psi(x, -y). \tag{F.8}$$

Lastly, we require that the stream function decay to zero at infinity:

$$\Psi(x, \pm\infty) = 0. \tag{F.9}$$

Our analysis begins with the unsteady Stokes equation

$$\partial_t \mathbf{u} = -\frac{1}{\rho}\nabla p + \nu\nabla^2 \mathbf{u}. \tag{F.10}$$

Taking the curl of both sides, we find, in terms of the stream function Ψ,

$$(\partial_t - \nu\nabla^2)\nabla^2\Psi = 0. \tag{F.11}$$

Analyzing just one Fourier component, $\Psi_{k\omega}(x, y)$, this equation takes the form

$$\left(\frac{i\omega}{\nu} + \nabla^2\right)\nabla^2\Psi_{k\omega} = 0. \tag{F.12}$$

We look for solutions of the form

$$\Psi_{k\omega}(x, y, t) = \psi_k(\omega, y)\, e^{i(kx-\omega t)} \tag{F.13}$$

and find, using conditions (F.8) and (F.9), that

$$\psi_k(\omega, y) = ae^{-k|y|} + be^{-q|y|}, \tag{F.14}$$

where

$$q = \sqrt{k^2 - i\omega/\nu}, \tag{F.15}$$

in which the square root is defined such that its real part is positive. Note that the jump condition (F.5) is automatically satisfied when equation (F.3) holds, since from (F.1) and (F.6),

$$S_{xy} = \rho\nu(-\partial_x^2\Psi + \partial_y^2\Psi) \tag{F.16}$$

is an even function of y. Since $\partial_y \psi(y)$ is odd and continuous, it vanishes at $y = 0$ and we have $b = -ka/q$. Thus

$$\psi_k(\omega, y) = a \left(e^{-k|y|} - \frac{k}{q} e^{-q|y|} \right). \tag{F.17}$$

The pressure may be obtained from the unsteady Stokes equation (F.10). Again assuming that the interface is located at $y = 0$, we have

$$\partial_y p = -\rho(-\partial_t + \nu \nabla^2)\partial_x \Psi. \tag{F.18}$$

Analyzing just one Fourier component as before, we obtain

$$\partial_y p_k(\omega) = -ik\rho(i\omega + \nu \nabla^2)\psi_k(\omega). \tag{F.19}$$

Now note that, using the definition (F.15) of q,

$$(i\omega + \nu \nabla^2)\psi_k(\omega) = i\omega a e^{-k|y|}. \tag{F.20}$$

Thus, from (F.19), we have

$$\partial_y p_k(\omega) = k\rho\omega e^{-k|y|}. \tag{F.21}$$

Integrating over y and requiring that $p_k(\omega, |y| \to \infty) = 0$, we obtain

$$p_k(\omega, y) = -\text{sgn}(y)\rho\omega a e^{-k|y|}. \tag{F.22}$$

We thus find that the jump in the pressure is

$$[p_k(\omega)] = -2\rho\omega a. \tag{F.23}$$

Then from (F.4) and (F.6) we have that

$$[S_{yy}(x, t)] = -[p(x, t)]. \tag{F.24}$$

Thus, from equations (F.7), (F.23), and (F.24), we find

$$a = -F_0/(2\rho\omega). \tag{F.25}$$

Now note that when the velocity u_y is evaluated at the interface it is equal to the interface height velocity $\dot{H}(x, t)$. It may be obtained by evaluating the stream function (F.1) at the approximate location of the interface, $y = 0$. Using (F.13) and (F.17) to express the height velocity of a single Fourier component, we obtain

$$\hat{\dot{H}}_k(\omega) = -ik\psi_k(\omega, y = 0). \tag{F.26}$$

Substituting equation (F.25), we find

$$\hat{\dot{H}}_k(\omega) = \frac{ikF_0}{2\rho\omega} \left(1 - \frac{k}{q} \right). \tag{F.27}$$

This equation takes the form

$$\hat{\dot{H}}_k(\omega) = \hat{R}_k(\omega)F_0, \tag{F.28}$$

where $\hat{R}_k(\omega)$ is the hydrodynamic response function given by

$$\hat{R}_k(\omega) = \frac{ik(1 - k/q)}{2\rho\omega}.$$

(F.29)

Appendix G
Answers to exercises

2.4. $\mathbf{v} = B\mathbf{u}$, $\omega' = \omega$.

5.3. 32.

9.1. First, note that in the 14-bit ILG there are $3^7 = 2187$ distinct configurations. Then use a computer (or think carefully) to determine the number of unique color-flux vectors $\mathbf{q}$ that these 2187 configurations define. Somewhat more effort then gives the number of distinct unit vectors $\hat{\mathbf{f}}$ that are defined by the $15^6 \approx 1.1 \times 10^7$ possible neighboring color distributions $(\phi_i)_{i=1,\dots,6}$. Then straightforward but tedious computation supplies the desired answer. We have never tried this ourselves, but a better understanding of the approximation $\hat{\mathbf{f}} \approx f_*$ could be important for future implementations on special-purpose computers with limited memory capacity.

11.2. One way to implement the dumbbell method would be to first locate dumbbells and empty pairs by Boolean logic and to also count the number of dumbbells $n_d = \sum_i d_i$ (an arithmetic operation which could itself be an entry to a lookup table). Then a lookup table would take as input

1. a 12-bit entry in which 1's indicate paired directions where dumbbells can be placed;

2. n_d, at most a 4-bit number, indicating how many dumbbells must be placed; and

3. f_*, a 2-bit code indicating the archetypal gradient.

In such a form the table would have 2^{18} entries.

11.3. As we see in Chapters 2, 4, and Appendix A, anisotropy disappears at a macroscopic scale for any tensors constructed from lattices with sufficient symmetry. However, surface tension is not a tensor and is therefore not expected to inherit any such symmetries. As a practical matter, one need only consider 2D Boolean ILG collisions with different interface orientations to perceive the anisotropy.

Author Index

Subject Index

{3,4,3} polytope, 268

advection, 16
anisotropic, 35
automata, 168

Bénard-von Kármán streets, 4
binary fluid, 204
biological systems, 5
biology, 151
Boltzmann approximation, 46
  definition, 186
  immiscible lattice gases, 208
Boltzmann equation, 48, 58
Boolean analog computer, 8
Boolean analog fluid, 9
Boolean dynamics, 184
Boolean variables, 21, 182
boundary conditions, 86–89
Brownian motion, 238, 246
bubble drag, 137–138

Cahn-Hilliard free energy, 205
canonical ensemble, 177
capillary number, 124
capillary waves, 126, 229, 230, 237
cellular automata, 5–6
  definition, 168
cellular automata machines, 6
CFL number, 84–85
Chapman-Enskog expansion, 47,
  187–196, 201
chemical potential, 36, 178, 205
collision operator, 22, 30, 169

FCHC, 67–70
  linearized, 48–49, 52–53, 80, 95
  pseudo-equilibrium, 74
collisions, 2
color
  conservation of, 92
  field, 108
  flux, 108
  gradient, 108
complex fluids, 7, 239
compressible, 17, 271
computing spaces, 5
configuration
  global, 35, 171
  local, 35, 168, 181
contact angle, 158
continuum mechanics, 3, 12–14
convection-diffusion equation, 93–94
critical temperature, 204

Darcy's law, 153, 165
  two fluids, 154, 155
de Bruijn diagram, 168, 169, 171
detailed balance, 170
diffusion, 93, 206
  coefficient, 94–97
    measurement, 96
  immiscible lattice gases, 207
dispersion, 98–103
domain growth, 209
dumbbell method, 128, 130–131
dynamical systems, 6
  dissipative, 210, 218